SPECTROSCOPIC ANALYSIS OF GAS MIXTURES

SPECTROSCOPIC ANALYSIS OF GAS MIXTURES

O. P. BOCHKOVA And E. YA. SHREYDER
LENINGRAD STATE UNIVERSITY

Edited by
S. E. FRISCH
CORRESPONDING MEMBER
ACADEMY OF SCIENCES OF THE USSR

Translated by Scripta Technica, Inc.

TRANSLATION EDITOR
CHARLES P. POOLE, JR.
DEPARTMENT OF PHYSICS AND ASTRONOMY
UNIVERSITY OF SOUTH CAROLINA
COLUMBIA, SOUTH CAROLINA

1965

ACADEMIC PRESS New York and London

ACADEMIC PRESS INC.
111 Fifth Avenue, New York, New York 10003

United Kingdom Edition published by
ACADEMIC PRESS INC. (LONDON) LTD.
Berkeley Square House, London W.1

LIBRARY OF CONGRESS CATALOG CARD NUMBER: *65-26037*

Originally Published as:
"Spektral'nyy Analiz Gazovykh Smesey,"
Second, Revised and Augmented Edition, by Fizmatgiz, Moscow, 1963

PRINTED IN THE UNITED STATES OF AMERICA.

EDITOR'S FOREWORD

Spectral analysis of gas mixtures is a relatively recent addition to generally available analytical techniques, and presents its own peculiarities, as well as some specific difficulties. In emission analysis, the main difficulties are connected with the excitation of spectral lines in a low-pressure gas discharge plasma and the inevitable evolution of gases (as well as their absorption) by the discharge tube walls. This has a negative effect on analytical accuracy. Nevertheless, the composition of gaseous mixtures can be successfully determined by spectral methods, and problems are now solved simply and rapidly by such techniques. Thus, spectral analysis is used for quality control of inert gases; this analysis is also used to test gases given off by various processes, to test air, etc.

Analysis of gas mixtures is less sensitive than that of metals. At best, the limit of detection is not lower than 10^{-4} %. Increasing this sensitivity is a difficult problem, which will be solved eventually. On the other hand, the absolute limit of detection of this analysis may be higher in the case of gases ($\sim 10^{-10}$ g) than for metals. The spectra of most gases show rather widely spaced lines, or very distinct molecular bands. Because of this, fast-response photoelectric gas analyzers using light filters can be designed. In these, the dispersing system can be eliminated. Such equipment should prove useful in solving numerous analytical problems.

In spectral gas analysis, generally applicable techniques cannot be devised as readily as in cases where metals or alloys are analyzed. The problems involved for the most part require specific solutions. Consequently, there is a much-felt need for a pertinent manual—all the more so in view of the paucity of available literature on the subject. "Spectral Analysis of Gas Mixtures," by O. P. Bochkova and E. Ya. Schreider, published in 1955, has been out of print for some time. Thus a new edition was in order.

In the eight years that have elapsed since the date of the first edition, an impressive number of papers on spectral gas analysis were published both in the Soviet Union and abroad. New techniques have been developed. The range of problems that can be solved has been broadened. This called for a revision of the book, with the incorporation of recent material. At the same time the authors, not wishing to expand the work too far, deleted the general discussion of spectroscopic and spectral-analytical techniques, since this is a topic treated in generally available books.

The new edition offers a more extensive description of photoelectric procedures, which are now widely used by spectroscopists. Some recently developed quantitative methods of fast gas analysis are described. As before, the book is concerned primarily with emission analysis. To render it more comprehensive, however, Chapter VI on absorption analysis was materially extended to include spectroscopic procedures based on absorption in the ultraviolet and infrared regions of the spectrum, the optico-acoustic method, etc.

The appended tables help in selecting specific conditions for analyzing the most commonly used mixtures. Reproductions of spectrograms are given for most of the gases dealt with in practice. Ample references will further aid the researcher in solving problems of a more specific nature.

S. E. Frisch

CONTENTS

FOREWORD

This handbook is concerned specifically with methods applied in spectral analysis of gas mixtures. Other methods of gas analysis have been described in general manuals [1-4].

The work is based in part on data obtained in the Optics Laboratory of the Research Institute of Physics, Leningrad University, during experimental studies designed to develop new techniques for gas mixture analysis. The experiments were conducted in collaboration with L. P. Razumovskaya and N. V. Chernysheva, under the supervision of Prof. S. E. Frisch, Associate Member of the Academy of Sciences of the USSR.

The authors are greatly indebted to S. E. Frisch, who edited the book, and to L. S. Lomonosova for her valuable suggestions and comments offered during the reading and discussion of the manuscript. Further thanks are due to copy editor L. I. Orlova, who accomplished the laborious task of preparing the manuscript for publication.

O. Bochkova
E. Schreider

Chapter I

EMISSION FROM GASEOUS DISCHARGES

1. EXCITATION OF SPECTRA

The spectra of monoatomic gases consist of separate spectral lines. These appear when the atom passes from a higher to a lower energy level.

In spectroscopy the atomic energy levels are frequently plotted as horizontal straight lines, each line representing a different level, with the energy of the atom in its ground state taken as the zero reading. The energy levels are usually given in electron volts and the frequencies in reciprocal centimeters. The simplest energy level diagram, that for the hydrogen atom, is shown in Fig. 1. The diagrams for other elements are far more complex. Thus, Fig. 2 shows the singlet and triplet energy levels of the helium atom.

Wavelength tables and special atlases [5-9] have been compiled to facilitate spectra identification. Energy level graphs for various elements, or data from which such charts can be plotted, may be found in multiplet tables [10-14].

In the conventional classification adopted in spectroscopy, an energy level is identified by the quantum numbers that specify it [15-19]. A spectral line, corresponding to a transition between levels, is denoted by two groups of symbols, the first relating to the lower and the second to the upper energy level. Thus, the helium line with λ = 3888.6 Å is identified as $1s\,2s^3S_1 - 1s\,3p^3P_2$.

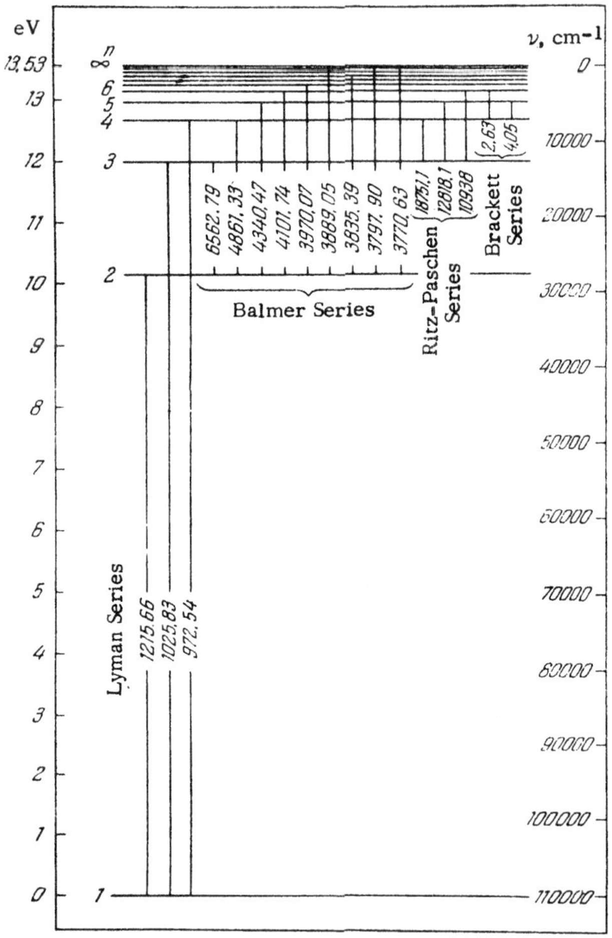

FIG. 1. Energy level diagram for the hydrogen atom.

Not every transition between levels is possible since the allowed transitions are limited by selection rules. Thus the helium atom (see Fig. 2) cannot exhibit transitions from the $1s\,2s\,^1S_0$ and $1s\,2s\,^3S_1$ states to a lower level. States from which a transition to a lower energy state is forbidden by selection rules are called *metastable*. A metastable state has a longer lifetime

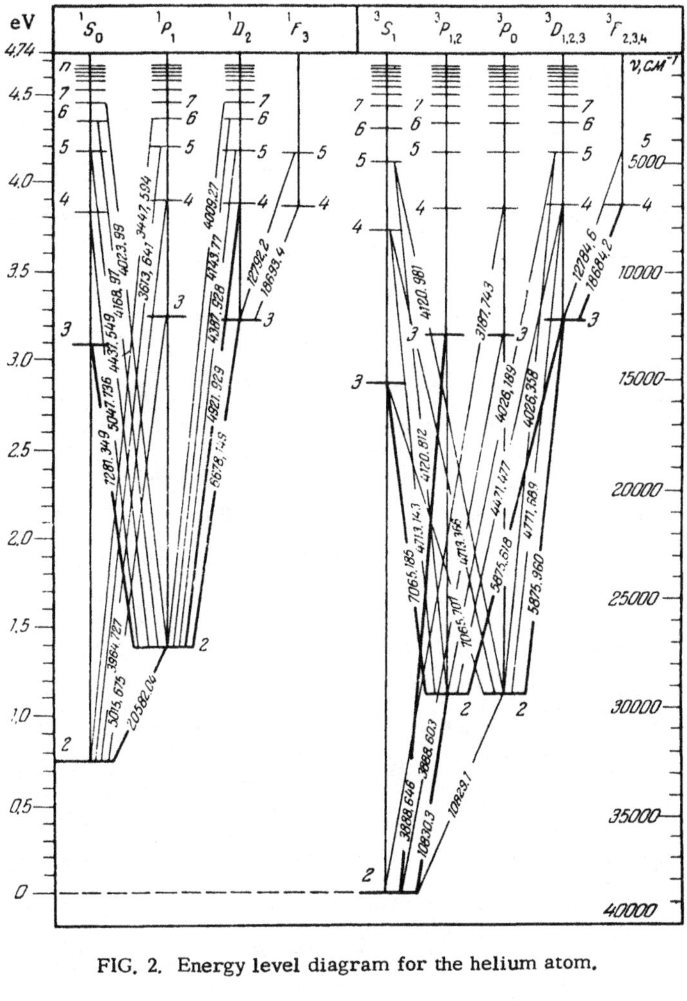

FIG. 2. Energy level diagram for the helium atom.

than other states. While the average life of a radiating level is 10^{-8} seconds, those of metastable levels may be of the order of 10^{-1} seconds.

Under ordinary conditions, the return of an atom from a metastable to a ground (normal) level may occur in one of two ways: 1) an atom, colliding with an electron, is raised to an excited state, from which it can pass to the ground state, the transition being accompanied by radiation; or 2) the excited atom transfers its energy to another atom (*collisions of the second kind*) and is reduced to the ground state without emission of radiation.

The spectra of other inert gases (neon, argon, krypton, xenon) are structually more complex than that of helium. This is due to the greater number of electrons in the outer shells of these atoms; thus, each outer shell contains two s electrons and six p electrons. An energy level diagram for the neon atom is shown in Fig. 3. Its ground level—$1s^2\,2s^2p^6S_0$—is very low. Lying considerably higher (by 16.5-16.8 eV) are the closely spaced $^3P_2$, $^3P_1$, $^3P_0$ and $^1P_1$ levels which correspond to the electron configuration $1s^22s^22p^53s$. Two of these states, $^3P_0$ and $^3P_2$, are metastable. As the neon atom returns from levels $^3P_1$ and $^1P_1$ to the ground state $^1S_0$ it emits two resonance lines in the far UV region with λ = 744 and 736 Å, respectively. Above these $2p^53s$ levels is located a group of ten states in the energy range of 18.3-18.9 eV. In the Russell-Saunders chart these levels are designated as $^3S_1$, $^3D_3$, $^3D_2$, $^3D_1$, $^1D_2$, $^1P_1$, $^3P_2$, $^3P_0$, $^3P_1$ and $^1S_0$ (electron configuration $1s^2\,2s^2\,2p^5\,3p$). The transition from these levels to lower ones produces a group of high-intensity lines observed in both emission and absorption spectra of neon excited in a gaseous discharge. On the other hand, the transition of the neon atom from higher states to the levels associated with this $1s^2\,2s^2\,2p^5 3p$ configuration produces lines in the infrared, red and orange regions of the spectrum. The neon ionization potential is 21.6 V. The other inert gases (argon, krypton and xenon) exhibit similar spectra. The greater the molecular weight of the inert gas, the closer the various energy

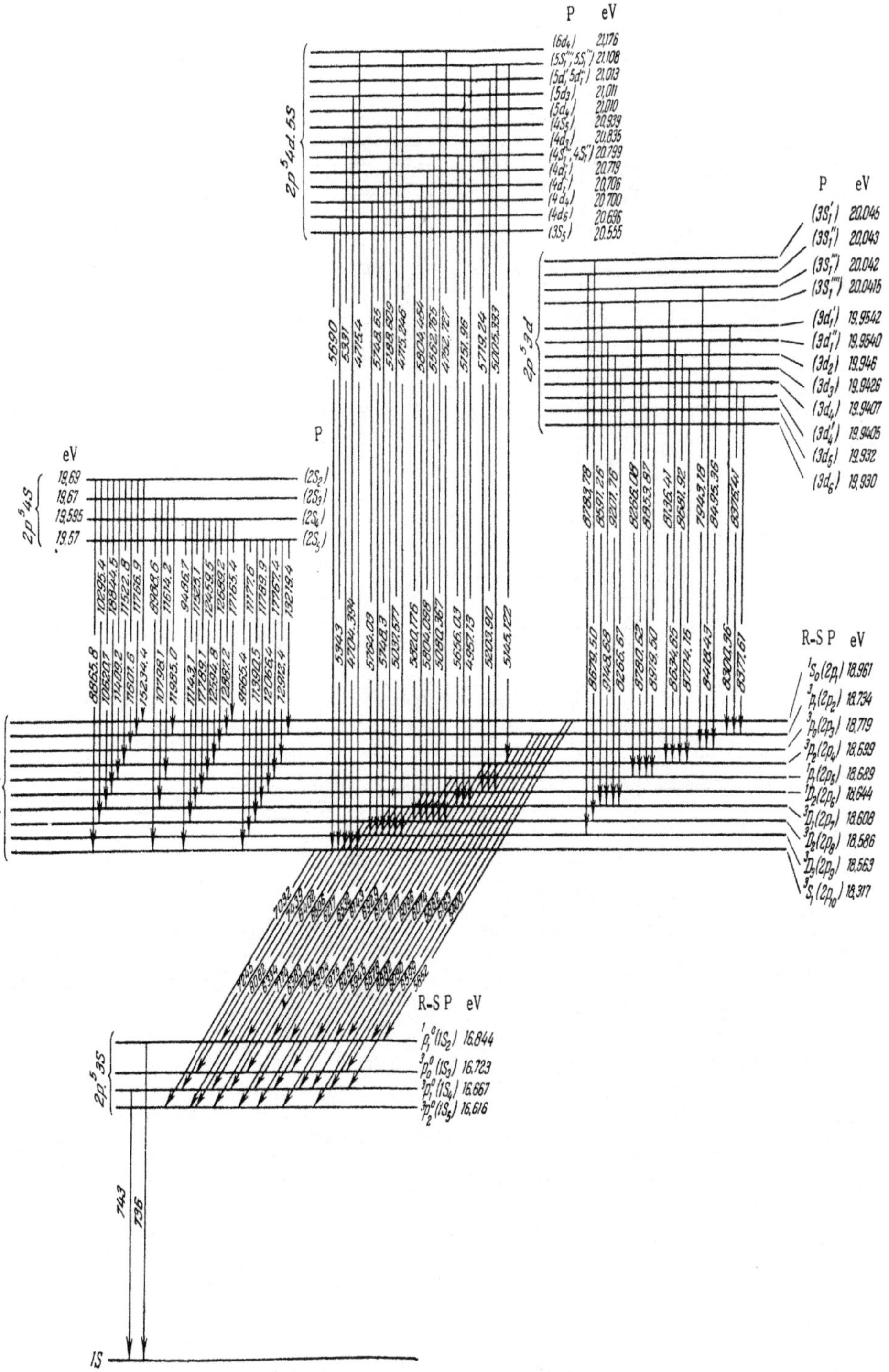

FIG. 3. Energy level diagram for the neon atom. R-S—Russell-Saunders chart; P—Paschen chart.

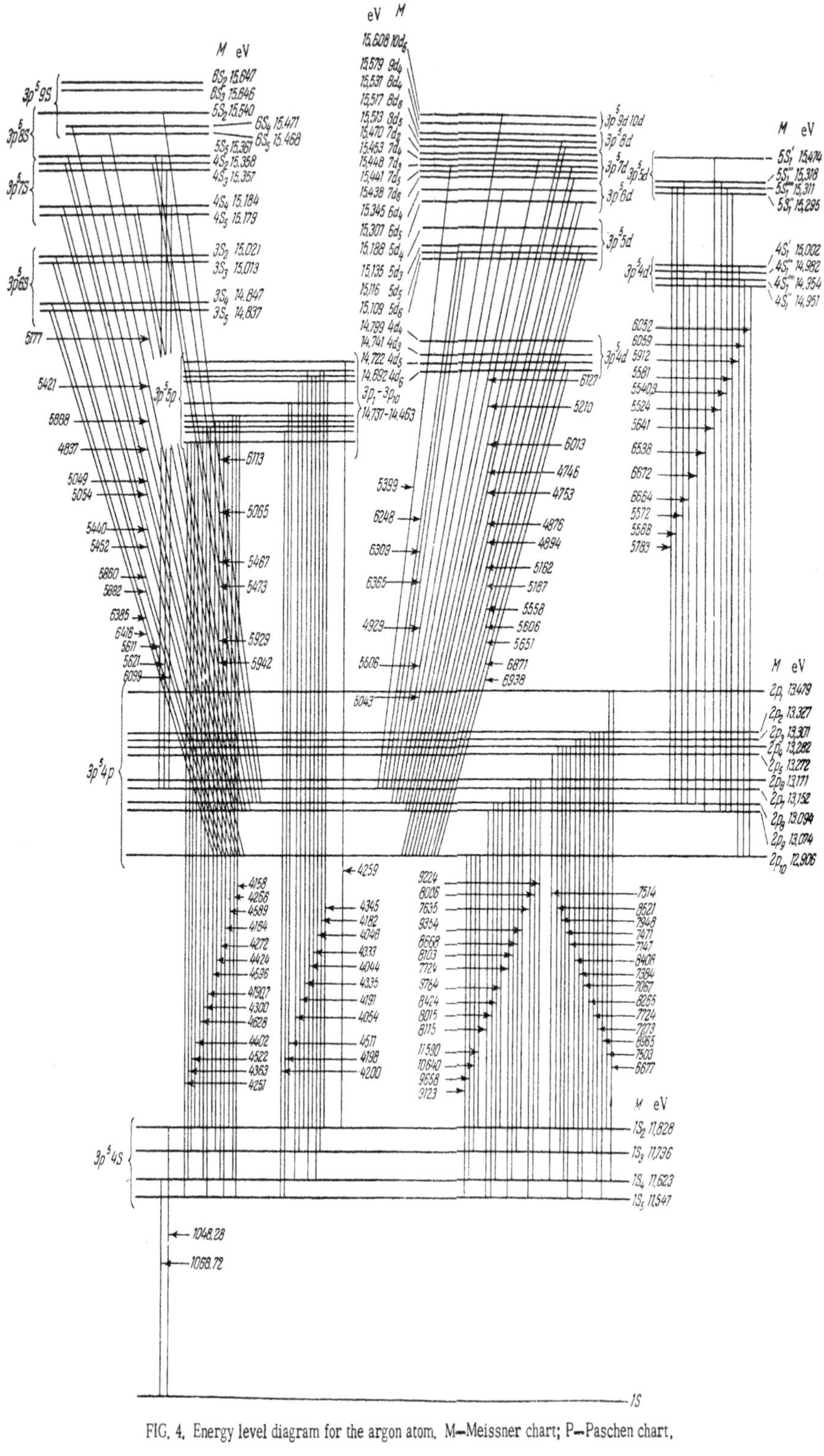

FIG. 4. Energy level diagram for the argon atom. M—Meissner chart; P—Paschen chart.

levels and the more marked the displacement of the entire spectrum toward long wavelengths (see, for example, Fig. 4).

Molecular spectra may also used in analyzing gas mixtures [16, 19]. Thus, diatomic and polyatomic molecules produce band spectra and gas analysis is usually based on the intensities of band edges, whose wavelengths are known for most gases and can be looked up in reference books [20]. Photographs of spectra of the inert gases and some molecular gases are given in Appendix VI.

The analysis of gas mixtures may employ both emission and absorption spectra.

The excitation of an atom [21-25] requires an energy greater than or equal to the excitation energy of a given level. This energy can be supplied to the atom by 1) conversion of the kinetic energy of electrons, ions or atoms impinging on the atom to excitation energy; 2) absorption of light quanta; 3) collisions of the second kind, that is, radiationless transfer of excitation energy from other atoms. In the last case, the efficiency of transfer is highest if the energy given up by the donor atom approaches the excitation energy of the given state of the recipient atom.

Let us now consider the first of the above methods of excitation, assuming an electron to be the impinging particle. Such collisions may be either 1) elastic, with the electron recoiling and almost no transfer of electron energy to the atom (the energy is distributed between the colliding particles in proportion to their masses); or 2) inelastic, with the electron transferring some or all of its energy to the atom, as a result of which the atom becomes excited. An electron—atom collision can become inelastic only if the electron possesses sufficient energy to raise the atom to an excited state at least one level higher than the ground state. Not all collisions will produce excitation, however—not even all those that involve electrons with sufficient excitation energy. This is because there is a definite probability of energy transfer on collision. The *probability of excitation* is expressed as the ratio of the number of inelastic collisions to the total number of collisions with electrons of a given velocity. It should be noted at this point that an electron—atom collision should not be regarded as a mechanical collision of two bodies; consequently, the number of collisions between the electron and the atom will be determined not by the dimensions of the latter (as given by the kinetic theory of gases) but by the *effective cross section of the atom* $Q = \pi R^2$, where R, the range of action of the atom, is the maximum distance (from the center) at which the electron is still able to transfer its kinetic energy to the atom.

The dependence of the effective cross section of the atom on the velocity of the impinging electrons may be expressed by the excitation function. We distinguish between the *level excitation function*, which describes the dependence of the effective cross section on the electron velocity relative to the excitation of a given level, and the *spectral line excitation function* (the optical excitation function), which expresses the dependence of the effective cross section on the electron velocity relative to the excitation of a given line. The spectral line excitation function differs from the level excitation function in cases where successive transitions are of importance, i.e., in cases when the atom undergoes transitions from all the higher states to a given level.

An exhaustive discussion of the phenomena associated with electronic and ionic phenomena will be found in the book by Massey and Burhop [26].

Because of the experimental difficulties involved, excitation functions have been determined only for a few elements. They were most extensively investigated for helium [27-31], mercury

[32, 33] and neon [34, 35]. Most of the existing measurements are relative, with few absolute values available.

Schaffernicht [32], Frisch and Zapesochniy [33], and Smit and Jongerius [36] obtained excitation functions with two or more peak values. It was shown that in most cases additional maxima emerge as a result of successive transitions [33]. In some instances secondary maxima also appear in the excitation functions of individual levels [27].

Excitation function maxima, as demonstrated by many investigators, are sharper for triplet than for singlet lines, and are located near the excitation potential. Single line maxima are shifted toward high velocities.

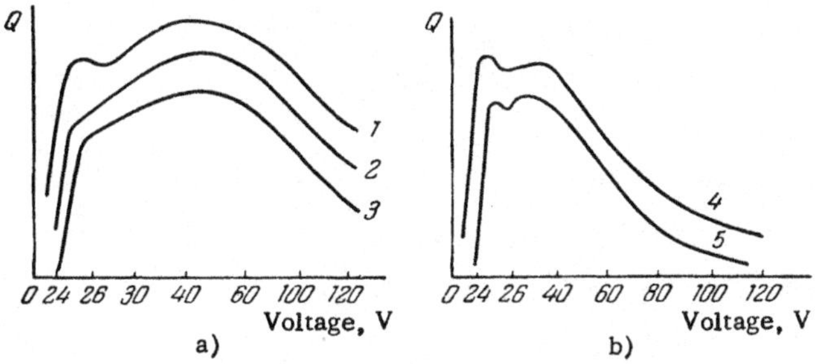

FIG. 5. Excitation functions of singlet (a) and triplet (b) helium lines. 1—λ4921 Å, 2—λ4388 Å, 3—λ4143 Å, 4—λ4713 Å, 5—λ4121 Å

Figure 5 shows excitation functions of five helium lines [27]. The shape of the function is determined by specific characteristics of a given energy level such as its multiplicity. If the electron velocities are lower than those corresponding to the level excitation potential, the effective cross section is zero, since electronic impacts will be limited to elastic collisions. An increase in these velocities is reflected in a rapidly increasing excitation function, whose maximum lies at an electron velocity somewhat greater than that corresponding to the excitation potential of a given level.

The difference in behavior of the singlet and triplet excitation functions accounts for variations in the intensity ratio of two spectral lines of different multiplicity when the pressure is changed. The gas pressure determines the energy of the electrons. The energy level is high at low pressure, and therefore singlet lines predominate. High pressures mean relatively low electron velocities; hence triplet lines predominate since their maxima lie at lower electron velocities.

A number of approximations for description of the behavior of the excitation function are available. Thus, Fabrikant suggested the following empirical expression [37, 38]:

$$Q_V = Q(V_m)\frac{V - V_A}{V_m - V_A}e^{1-\frac{V-V_A}{V_m-V_A}} \tag{1.1}$$

where $Q(V_m)$ is the peak value of the excitation function, Q_V is its value corresponding to a given energy level of the electron, V_A is the excitation potential, and V_m is the value of the potential corresponding to the peak value of the function. All the quantities appearing on the right side of Eq. (1.1) can be determined experimentally. Equation (1.1) correctly describes the behavior of an excitation function near the maximum for both singlet and triplet levels. Significant deviations appear only beyond the peak of the excitation curve.

A closer agreement with experimental values is given by the equation

$$Q_V = Q(V_m)\frac{V - V_A}{V_m - V_A}e^{1-\frac{V-V_A}{V_m-V_A}} + Q_\infty, \tag{1.2}$$

where the term Q_∞ is a constant [39].

If the values of the excitation function are known, then the absolute intensity of a spectral line can be calculated from

Eqs. (1.1) and (1.2). These expressions can then also be used to identify the specific factors affecting line intensity [37, 39, 40].

The shape of an excitation function often explains the variation of individual line intensities when the conditions prevailing in the discharge are changed. Thus in studies of pulse discharges by the method of time-resolved spectra the variation of intensity of a spectral line with time resembles its excitation function [41]; this is because the electron temperature (and hence the average electron velocity) varies during the pulse interval. For that reason the time resolution of a spectrum shows the intensity variation of a spectral line as a function of the electron velocity. In effect, we shall obtain an excitation function provided the electron beam is not monochromatic.

For analytical purposes the single lines, with their characteristic broad peak excitation functions, are most useful [42].

While it is now possible to calculate the effective cross sections of excitation for the simpler atoms [44-46], a satisfactory agreement between theory and experiment is not always attained.

The excitation function of an atom is not only affected by collisions with electrons, but also by collisions with ions [47]. While ionization and excitation brought about by collisions of the atom with ions are observed only at high velocities (most often at energies exceeding 1000 eV), partial ionization of atoms may occur at relatively low velocities [48]. Several papers report on spectra obtained by excitation with α-particles [49, 50].

It is also necessary to know the *ionization function*, i.e., the dependence of the ionization cross section on the electron velocity. The probabilities of ionization of various inert gases were reported in [43]. Ionization functions increase with electron velocities over the energy range of 100 to 150 eV, and then gradually decline.

The excitation functions for ionic energy levels are similar to those representing energy states of the atom. Ionic lines form the so-called *spark spectrum*, while atomic lines constitute the *arc spectrum* of an element.*

Spectral lines are produced not only by singly ionized atoms, but also by those which are multiply ionized. The energies required to obtain spectra of the latter [51, 52] are high and the spectra usually appear in pulse discharges.

Spectral lines are associated not only with direct excitation of atoms, but also with secondary processes such as *stepwise excitation* and collisions of the second kind. An excited atom may revert to its ground state (or undergo transition to a lower level) without emitting light provided it can transfer its excitation energy to an electron or another particle present during the discharge. Stepwise excitation, on the other hand, raises the excited atom to a still higher energy state. This is possible because the atom remains in an excited state for a finite time and thus may eventually undergo an inelastic collision with an electron, as a result of which it is raised to a state of higher excitation. Thus, the atom may store energy via several successive collisions with electrons. Stepwise excitation is sometimes quite important even though the probability of a collision between an excited atom and an electron is low. This is because stepwise excitation is favored by the presence of metastable atoms and diffusion of *resonance radiation.*** The theory of diffusion of resonance radiation was developed by Compton [53, 54] and refined by Biberman [55] and Holstein [56].

An excited atom will emit, on the average, one photon over its lifetime τ (prior to reverting to its ground state). This photon

*These names reflect the fact that ionic lines are excited primarily in a spark discharge and atomic lines in an arc.

**Resonance radiation is that accompanying transitions from the first excited level to the ground state.

may be absorbed by another atom in the discharge and will raise that atom to an excited state. The second atom will reradiate the photon upon reversion to ground state and may in turn excite a third atom, etc. The photon (or quantum of light) will thus be transferred from one atom to another. As a result, even though each individual atom remains in an excited state for only a short time, the probability that some excited atom will collide with an electron increases as if the life of the excited state of an individual atom were extended.

It was shown in theory and confirmed by experiment that in many instances a stepwise transition is more likely than a direct one [57-60].*

Stepwise excitation is a major factor in generation of spark spectra. Direct excitation of spark lines is possible only at high electron velocities (e.g., at low pressures) [62], since the kinetic energy of the impinging electron must be greater than the combined ionization and excitation energies of a givel level. Stepwise excitation of spark lines, on the other hand, is possible at low electron velocities. The energy of the electrons need only be sufficient to excite an atom. Then, on its second collision, the excited atom becomes ionized, while the third collision results in the appearance of an excited ion. However, stepwise excitation requires a high concentration of electrons [63].

Emission processes, as seen from the above discussion, are closely related to the life of excited atoms: the probability of secondary processes increases with the lifetime of an excited atom. Hence the importance of metastable atoms, for which the probability of stepwise ionization or collisions of the second kind is higher than for transitions accompanied by emission of light.

---

*Absolute values of effective cross sections for the stepwise excitation of neon are given by Frisch and Rewald [61].

The life of a metastable state depends strongly on the electron concentration and the presence of impurities whose excitation potentials are close to the excitation energy of the metastable level [64]. The metastable atoms can transfer their energy by collision with the atoms of the impurity, the overall result being a shorter lifetime of the metastable level.

The presence and the number of metastable atoms determine the conditions under which a self-sustaining discharge can occur. Thus, when the discharge gap is illuminated by an outside source, or specific (and effective) admixtures are added to the gas, the number of metastable atoms decreases drastically and the discharge firing potential increases [65-68].

The various processes occurring in the light source (excitation as well as annihilation of the excited atoms) finally result in a fixed concentration of excited atoms and ions within that source. This, in addition to other factors, determines the intensities of the spectral lines.

Another essential factor influencing the line intensity is *reabsorption*. The term refers to the fact that the radiation emitted by one volume element of the light source is reabsorbed to some extent in the other regions of the source prior to being radiated to the outside.

2. BASIC CHARACTERISTICS OF GASEOUS DISCHARGE

The occurrence of processes in a discharge arc is determined by the concentrations of electrons, ground-state and excited atoms, and the particle velocity distribution. The mixture of electrons, ions and atoms present in the gas discharge gap forms the so-called plasma. Plasma may be isothermal and nonisothermal.

An *isothermal* plasma exists in a state approaching thermodynamic equilibrium. It has a characteristic temperature T which determines the degree of ionization of the plasma material (the ion and electron concentration), the particle velocity distribution, and the distribution of particles over energy levels. These distributions are found from the familiar statistical equations of Saha, Maxwell and Boltzmann, and the temperature T appears in each of these expressions.

The average kinetic energy of electrons, ions and atoms is the same, and consequently the electron gas temperature is the same as that of gas itself. Such equilibrium can be attained at relatively high pressures, i.e., under conditions associated with a large number of collisions per unit time and an intensive energy exchange. One example of an isothermal plasma is an arc and spark generated at atmospheric pressure. Thermodynamic equilibrium, however, will not always establish itself even at this pressure level. For instance, it is not reached in arcs struck in inert gases [69, 70].

In a nonisothermal plasma, atoms, electrons and ions are all at different energy levels.

Electron temperature and electron concentration

The average kinetic energy of plasma electrons can be related to a specific temperature, generally referred to as the *electron temperature* (T_e). The term is merely a convention since thermodynamic equilibrium does not exist in a nonisothermal plasma and therefore the concept of temperature, in the accepted sense, becomes meaningless.

Electrons, as a rule, have greater energies than the average energy of the heavier particles. The reason for this is that electrons involved in elastic collisions with the heavy particles (atoms or ions) are able to transfer to the latter only an insignificant fraction of their energy, the difference in mass between the colliding particles being very large. Substantial energy transfers are possible only in the case of inelastic collisions, which account for only a minor fraction of the total number of collisions.

The electron temperature in the positive column (see p. 33) may be as high as a few hundred thousand degrees; the ion temperature does not exceed 1500–2000°K; and the average gas temperature is of the order of 500–1000°K.

The electron velocity distribution patterns obtained in numerous experiments tend to follow closely the Maxwellian distribution curve [71, 72]. At the same time, marked departures from this law were noted even at low pressures and weak currents [73, 74, 74a]. Such deviations are of particular importance in studying spectral line intensities, since excitation is largely associated with the tail portion of the Maxwellian curve (inelastic collisions).

In the case where the distribution of electron velocities follows Maxwell's law the electron temperature is the basic parameter of the discharge, controlling the conditions for the excitation of the spectrum, i.e., the spectral line intensities.

The other characteristic of the discharge which is of great importance is the *electron concentration* (n_e). It determines the number of exciting collisions, along with the probability of stepwise ionization and excitation processes.

Both electron temperature and electron concentration can be determined using the classic *probe method* developed by Langmuir [21, 22, 71]. The conditions necessary for the application of this method are discussed at length by Klyarfel'd [75].

At the present time, there exist various modifications of the probe method. Thus, a two-probe procedure has been developed [76–78]. Another recent technique exploits the ionic portion of the discharge characteristic [79]. In still another modification the probe data are recorded oscillographically [74, 80].

By means of probes, ignited with the aid of a high-power sweep oscillator, correct characteristics can be obtained within a short time [81, 82]. A procedure using a high-frequency resonance probe [83] is employed to obtain instantaneous values of electron concentration in unstable discharges. The probe method has found further application in studies of high-frequency discharges [84-87].

The electron temperature can also be measured by the noise method. Temperature measurements based on the probe and noise methods show good agreement.

Interferometric procedures [88, 89] are also effective in determining electron concentrations. These techniques, based on the fact that the refractive index of a plasma depends on the degree of ionization, are used in studies of pulse discharges to follow changes in electron concentration with time, as well as changes in the concentration gradient along the cross section of the tube.

Effect of discharge parameters on electron temperature

The electron temperature is a function of the current in the discharge gap [75, 90-95], the pressure [75, 92, 96-99], the discharge tube diameter [91, 92], and the composition of the luminescent mixture [75, 100-104]. Most of the references describe studies of discharges in pure gases.

The electron temperature drops with an increase in pressure because this produces a shorter mean free path. In consequence, an electron loses larger quantities of energy on collision and is unable to store much energy before such collisions. Electrons move slowly at high pressures. As the pressure increases, the electron temperature approaches the (kinetic) temperature of the gas and may be identical to it (isothermal plasma).

In molecular gases the electron temperature decreases with increasing pressure not only because the mean free path is reduced, but also because of changes in the degree of molecular dissociation. For dissociated molecules the ionization probability can be higher or lower than that for undissociated molecules. Dissociation may therefore cause an increase or a decrease in the value of T_e [98].

The electron temperature, as a rule, gradually decreases with an increase in the strength of the current. The effect is particularly marked at high pressures, i.e., under conditions where stepwise ionization becomes a major factor.

The electron temperature also depends on the diameter of the discharge tube, because an increase in the diameter causes a decrease in the potential gradient, and at constant pressures the electron temperature may be assumed roughly proportional to the potential gradient [105]. In small-bore capillaries the electron temperature may increase with the current due to rarefaction of the gas [94, 95].

The electron temperature depends on the type of gas that fills the discharge tube, and, other things being equal, it is greater for gases with higher ionization potentials [75]. Zaytsev, Dorgelo and others have shown that the electron temperature of a gas mixture decreases with an increase in the concentration of a readily excitable constituent. Zaytsev [101] analyzed a mixture of neon and argon at $p = 3$ mm Hg and $i = 7.5$ mA and obtained the following results:

Argon concentration, %	0	0.005	0.01	0.05	0.1	1
T_e, eV	3.2	3.16	3.11	3.07	2.95	2.43

Thus the discharge conditions in a gas mixture vary when the concentration of the easily excitable component is altered. However, addition of argon to neon will not necessarily decrease the electron temperature. Thus Penkin [106] reported that the addition of 0.5% argon to neon did not materially alter the electron temperature of a discharge proceeding at a pressure of 0.9 mm Hg and a current of 60 mA.

The data of Zaytsev and Penkin can be made to agree if we consider that the significance of additional ionization introduced by the atoms of the admixture increases as the $\frac{E}{p}$ ratio decreases (E is electric field strength; p is pressure). This is because at large $\frac{E}{p}$ the degree of ionization of a pure gas is already sufficiently high without additives [107]. We see that in Penkin's experiments the $\frac{E}{p}$ was more than an order of magnitude higher than in those of Zaytsev.

The electron temperature was markedly lowered when small amounts of krypton and xenon were added to neon [106]. The addition of a readily ionizable admixture reduced the number of ions of the species that is difficult to ionize [108], but the total number of ions increased due to collisions of the second kind and ionization of the admixed atoms by electron impact. Because of this additional ionization, the discharge could be sustained at a lower electron temperature.

A small amount of an admixture with an ionization potential higher than that of the basic component of the mixture does not materially alter the ionization conditions so that the electron temperature remains approximately constant. In gas mixtures exhibiting separation of the components (see Section 4), the electron temperature decreases along the positive column from the anode to the cathode. The component which is difficult to ionize glows primarily near the anode [103, 104].

Effect of discharge parameters on electron concentration

The electron concentration depends on the current density and electron temperature:

$$n_e = A \frac{jB}{\sqrt{T_e}} \tag{1.3}$$

where A is a constant, j is the current density, and B is the bypass factor.

The validity of this equation was tested by Klyarfel'd [75]. The *bypass factor* is the ratio of the electron path length to the displacement of this path in the direction of the field. The value of this factor is determined by the ratio of the densities of the random and directed electron currents. Klyarfel'd demonstrated experimentally that B increases with pressure at constant current and that the electron concentration also increases with pressure due, first, to the increase of B and, second, to decreased electron temperature. It further follows from Eq. (1.3) that the electron concentration increases more rapidly than does the current density, since the bypass factor increases, and the electron temperature decreases, with increasing current density. The linear relationship between current and electron concentration thus holds only to a first approximation, whereby the bypass factor and electron temperature variations are not taken into consideration.

Concentration of excited atoms and methods of determining it

The spectral line intensity depends on the concentration of excited atoms in the discharge. If no secondary processes are involved, the spectral line intensity is directly proportional to the concentration of atoms in the upper excitation level. The concentration of excited atoms in the lower level determines the value of reabsorption. The concentration of excited atoms determines the probability of collisions of the second kind and of stepwise transitions on excitation and ionization. Hence, the knowledge of concentration of excited atoms in the discharge is absolutely necessary for understanding of the emission mechanism. Various methods are available for determing this concentration.

1. *The "hook" method of Rozhdestvenskiy* [109, 110]. This procedure determines interferometrically the dependence of the refractive index on the wavelength in the neighborhood of the absorption line. This is the most accurate and reliable technique for determining the concentration of excited atoms, and has been repeatedly used for determining the concentration of excited atoms in vapors of metals and in inert gases [111-115].

2. *The absorption method* [24, 116-119]. The classic version of this method uses two identical discharge tubes, and measures the fractional intensity of the light from one tube after passage through the other. The concentration of excited atoms is then determined from the measured value of the absorption coefficient [24, 117]. This concentration can also be measured in those cases where the discharge conditions in the emitting and the absorbing tubes are not the same. It is then necessary to analyze the emission and absorption line patterns, and on this basis to determine the relationship between the absorption coefficient and the concentration of excited atoms. The problem thus becomes more complicated and can be solved only if precise data on both patterns are available.

3. *The reabsorption method* [37, 40, 120-126]. The electron population in any given level is determined by measuring the ratio of intensities of a given spectral line in two sectors of the luminous column. If the spectrograph is operated at full transmission capacity and no reabsorption is involved, the spectral line intensity is proportional to the length of the luminous column. The effect of reabsorption alters this relationship, which no longer depends on the ratio of the column lengths but is a single-valued function of the concentration of atoms in the lower level. A simplified variant

of this method uses a mirror, which magnifies the luminous column to double its length, set behind the tube (this variant does not allow for reflection losses).

4. *The spectral line width method* [127]. The excited atom concentration can be determined by measuring the width of a spectral line emitted by a light source possessing a finite optical density.

5. *The emission method* [128]. In the absence of non-radiative transitions and reabsorption of radiation, the concentration of excited atoms N_k is related to the spectral-line intensity I by the equation

$$I = N_k A_\nu h\nu, \tag{1.4}$$

where A_ν is the transition probability for a given line, h is Planck's constant, and ν is the radiation frequency.

6. *The reversal temperature method* [129-132]. In this procedure the discharge is transilluminated by a source with a continuous spectrum, operated at a variable temperature. One then ascertains the exact moment at which the spectral line, originating on the level in which it is desired to determine, vanishes against the source background. Knowing the temperature of the source, one can then calculate the *reversal temperature* and compute the ratio of concentrations of the excited atoms in the initial and final levels from the Boltzmann equation. To obtain the absolute values of these concentrations it is necessary to determine independently the concentration of excited atoms in either level.

Where no thermodynamic equilibrium exists, the reversal temperature varies for different lines. (The term "reversal" signifies that, for the given luminous volume and wavelength, there is a thermodynamic equilibrium between the radiant fluxes emitted by the continuous spectrum source and the source under study.) A simultaneous reversal of all lines and equal source and reversal temperatures indicate that a state of thermodynamic equilibrium exists throughout the system.

The first four methods give only the product of the concentration of excited atoms and the oscillator strength ($N_k f$). It follows that the numerical value of the concentration can be computed only if the oscillator strength is known. The last two methods are only suitable for determining the concentration of excited atoms at the radiating levels. The scope of application for each of the above procedures is given by Fabrikant [37].

Effect of discharge parameters on the concentration of excited atoms

The concentration of excited atoms depends on the current [113, 133-137]. Among other things it was demonstrated that the concentration of excited atoms may reach the saturation level at a certain current [133-137]. In the opinion of many authors this indicates an equilibrium between collisions of the first and second kind. Fabrikant [37] believes that the decrease of electron temperature with rising current is responsible for the saturation phenomenon. An increase in the current may have a twofold effect. On the one hand, the concentration of excited atoms should increase. On the other hand, this concentration should decline, due to the attendant electron temperature drop. These, however, do not appear to be the only causes of saturation under the conditions investigated by Fabrikant.

An essential factor affecting the concentration of excited atoms is its dependence on the electron velocity distribution. Departures from Maxwell's distribution may not appreciably affect the characteristic patterns obtained with the aid of probes, since these are determined largely by electrons whose velocities lie in the neighborhood of the maximum probability level. At the same time, the levels with high critical potentials are excited primarily by particles accounted for by the tail portion of the distribution function. For this reason, deviations from the Maxwellian-type electron velocity distribution may materially alter the process of level concentration of excited atoms as a result of an increase in current density may be explained by the lower electron temperature, as well as by the deviation of the velocity distribution of the perature, as well as by the deviation velocity distribution of the electrons from a Maxwellian-type curve [73, 99].

The decrease in the electron temperature was further shown by Kagan and Penkin [113] to be responsible for the absence of a direct linear relationship between concentrations of excited atoms and of electrons.

As the pressure increases, the concentration of excited atoms passes through a maximum, the lower energy states attaining that maximum at pressures which are lower than those necessary for the higher energy states [99]. The existence of a peak concentration is explained by the opposing effect of a pressure rise. Thus, while the concentration of excited atoms tends to increase because of higher concentrations of non-excited atoms and electrons, the higher pressures also tend to reduce the electron temperature which in turn effects a decrease in the concentration of excited atoms.

In Fig. 6 the variation of the concentration of excited atoms with pressure is shown for the case of a DC discharge in mercury vapor [113]. The shape of the curve is influenced by both primary and secondary processes [37]. Penkin and Shukhtin [131] and Kagan and Penkin [113] have shown that the mere fact that the electron velocity distribution is Maxwellian does not necessarily imply a Boltzmann distribution of atomic excitation levels. It is seen from

Fig. 6 that equilibrium in the $^3P_2$ level of mercury is reached only at relatively high pressures ($3 \cdot 10^{-2}$ mm Hg).

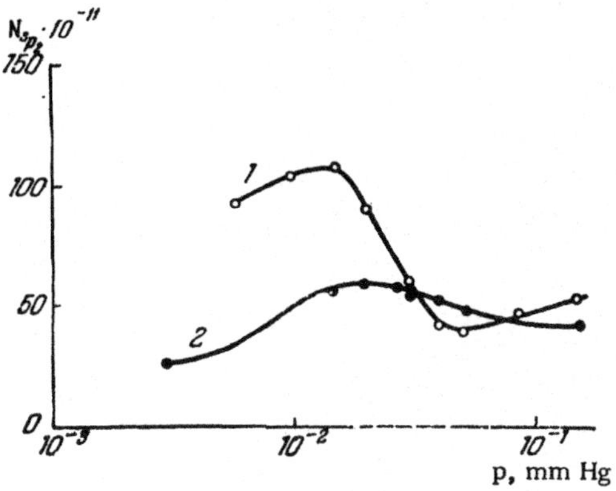

FIG. 6. Effect of pressure on the concentration of excited mercury atoms in the $^3P_2$ level. 1—Calculated curve based on the Boltzmann equation; 2—experimental curve.

At low pressures departures from equilibrium may occur even at vicinal levels. Thus, Bogdanova [138], in experiments with thallium vapor, observed an anomalous electron population in the $n^2D_{\frac{5}{2}}$ and $n^2D_{\frac{3}{2}}$ levels. Similar deviations from statistical equilibrium were noted for the hyperfine-structure components [139-141].

3. SPECTRAL LINE INTENSITY

All methods of quantitative spectral analysis are based on a comparison of the intensities of spectral lines. In spectral analysis of gases it is essential to understand the nature of the dependence of spectral line intensity on the current strength, the pressure and the concentration of elements in the mixture.* These

* We shall not consider such phenomena as reabsorption and collisions of the second kind, although these processes may significantly affect the excitation of a gas mixture, nor will we discuss the specific techniques used in measuring line intensities.

relationships then allow us to interpret the phenomena within the light source, since any change in discharge parameters does inevitably alter the intensities of spectral lines. The exact effects of discharge parameters on line intensities, as well as methods for calculating these intensities, are extremely complex problems which have been explored in many studies, notably those of Fabrikant [37] and Frisch [39, 40, 142].

Where no secondary processes are involved, the spectral line intensity is proportional to the number of excitation events $\Delta N_i(n_e)$ and to the quantum of energy $h\nu$:

$$I = K \Delta N_i(n_e) h\nu \tag{1.5}$$

where K is a proportionality factor that depends on the transition probability for a given spectral line.

In the absence of stepwise excitation, the number of excitation events can be calculated from

$$\Delta N_i(n_e) = n_e N_0 \int_{V_{0i}}^{\infty} Q_{0i}(V) F(V) \sqrt{V} dV; \tag{1.6}$$

where N_0 is the concentration of normal atoms, Q_{0i} is the effective cross section of the atoms, $F(V)$ is a function describing the electron energy distribution, and V_{0i} is the critical potential. Fabrikant substituted into Eq. (1.6) a value of $F(V)$ consistent with Maxwell's law and an excitation function calculated from Eq. (1.1). He thereby obtained a mathematical expression for the absolute spectral line intensity. We shall not show the full expanded equation but

$$I = K' f(T_e) n_e N_0, \tag{1.7}$$

where $f(T_e)$ is a monotonically increasing function of the electron temperature, and K' is a proportionality factor.

On the basis of (1.7) Fabrikant was able to explain the presence of intensity peaks in the spectral line at certain pressures. A higher pressure has a twofold effect. On the one hand, the concentrations of normal atoms and of electrons increase, and this in turn produces higher line intensities. This effect is, however, counteracted by the attendant electron temperature drop, which tends to lower the above intensities. It follows that the spectral line intensity may increase or decrease with a rise in pressure, depending on which of the two factors predominates. It is further evident that the lines which exhibit the maximum intensity at lower pressures are those with higher excitation potentials.

With stepwise excitation involving a single intermediate level, the number of excitation events is given by

$$\Delta N_{0i} = N_0 n_e \int_{V_{0i}}^{\infty} Q_{0i}(V) F(V) \sqrt{V} dV + $$

$$+ N_k n_e \int_{V_{ki}}^{\infty} Q_{ki}(V) F(V) \sqrt{V} dV, \qquad (1.8)$$

where Q_{ki} is the excitation function for the transition from level k to level i, V_{ki} is the critical potential for level i, and N_k is the concentration of excited atoms in the k state. A schematic representation of the excitational and radiative atomic transitions for the above case is shown in Fig. 7. Assuming $N_k \sim n_e$, the line intensity can be expressed as a sum of two quantities:

$$I = KN_0 [f(T_e) n_e + f_1 (T_e) n_e^2] \qquad (1.9)$$

where $f(T_e)$ and $f_1(T_e)$ are monotonically increasing functions of the electron temperature.

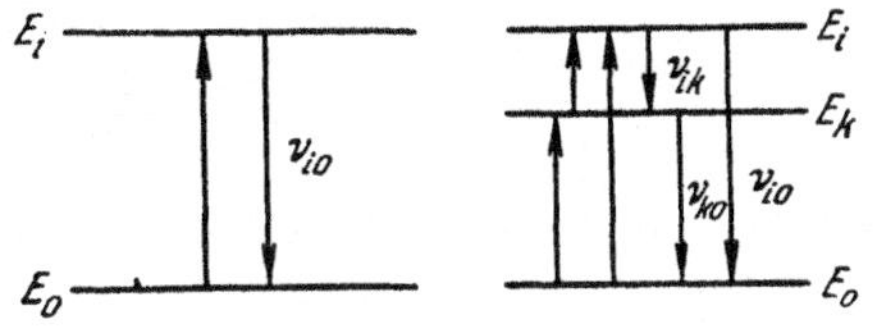

FIG. 7. Excitational and radioactive transitions in an atom.

As seen from (1.9) the spectral line intensity depends on both $f(T_e)$ and n_e. Higher currents can cause higher or lower line intensities depending on whether the electron temperature drop or the rise in the electron concentration proves controlling. Frisch [39] pointed out the possibility of a nonmonotonic variation of intensity following an increase in current density. Thus, over the range of very low current densities the intensity increases with electron concentration. This may be followed by a reduced intensity, due to a lower electron temperature. Finally, at large current densities, a higher intensity may again be noted since the phenomena corresponding to the second term of the equation can be important under these conditions.

At constant electron temperature and pressure, Eq. (1.9) reduces to the expression

$$I = An_e + Bn_e^2, \tag{1.10}$$

where A and B may be assumed constant.

The validity of this equation was confirmed by experiments of Frisch and Kagan [63]. At low current densities, the intensity of lines of neutral argon (Ar I) at first increased with the discharge current (see Fig. 8). The level sections of the curves, which are recorded at high currents, indicate the presence of secondary

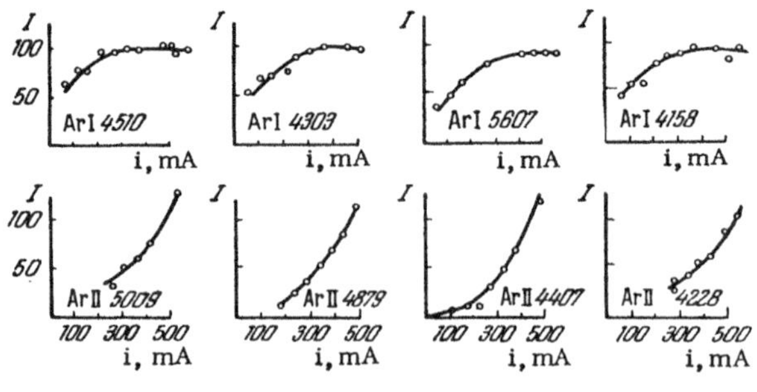

FIG. 8. Intensity of argon spark and arc lines vs. the current.

processes. The intensity of the lines of ionized argon (Ar II) varied as a square root of the current strength, indicating that the excitation is stepwise.

At high pressures (1–2 mm Hg), the curve of spark line intensities as a function of the current may exhibit a peak and a shape shown in Fig. 9 [94]. The nonmonotonicity of the curve is due to a transverse electric field along with a rarefaction in the central sectors of the positive column [94, 143, 144]. As the current increases, the pressure goes down. This may lead to lower line intensities.

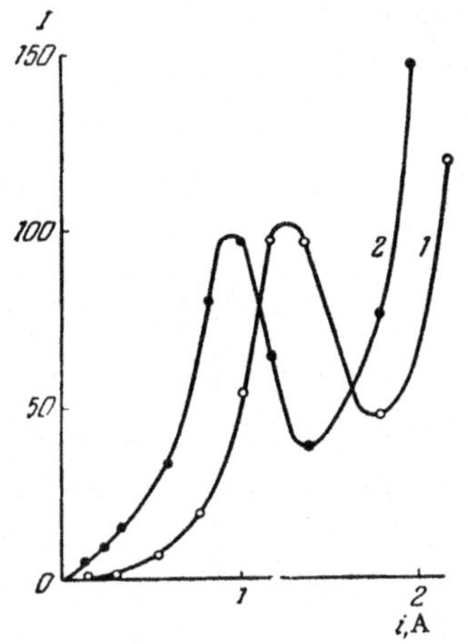

FIG. 9. Intensity of the argon spark line with $\lambda = 4806$ Å as a function of the current at pressures of 1.0 mm Hg (1) and 1.75 mm Hg (2).

The effect of the discharge parameters on the intensities of spectral lines in various light sources has been investigated in a series of experimental studies [145–151].

The above description of the effect of the discharge parameters on the spectral line intensity obviously is a simplified one. Actually, in calculating the intensity of radiation from a given spectral line, one should take into account possible deviations from the Maxwellian electron velocity distribution [73], in addition to various secondary processes such as successive transitions from higher to lower levels, collisions of the second kind, absorption of photons, recombination of ions, reabsorption of radiation, etc. [142].

To mention one instance, resonance radiation is quite important in the excitation of inert gases [27, 31, 152–155], and markedly distorts the spectral lines. Ladenburg and Levy [120] and Frisch and Bochkova [124, 125] studied reabsorption in neon while Frisch and Bogdanova [123] studied it in cesium and Harrison [126] in helium. In some instances reabsorption was noted even when the concentration of the impurity was low (1% oxygen in helium [126]). The reabsorption effect is not as pronounced when one inert gas is added to another [156]. In resonance lines, the reabsorption was observed even at very low pressures and low currents [152].

The effect of the various discharge parameters on the spectral line intensity is more difficult to establish in the case of gas mixtures than in pure gases. In this case we must allow for changes in line intensities caused by collisions of the second kind. It has been shown experimentally that such impacts may result either in intensification or weakening of the lines. Collisions of the second kind led to marked intensification of metal lines excited in an argon arc [157]. Molecular impurities intensified the mercury resonance line $\lambda = 2537$ Å [158]. The effect of collisions of the second kind on excitation in gas mixtures has been discussed in several reports [106, 142, 159-169].

For best analytical results it is advisable to work with a line whose upper excitation level has a brief lifespan, since a long life is associated with a high probability of collisions of the second kind [170].

The ratio of intensities of two spectral lines of various gases varies with the electron temperature. In a binary mixture, the component with the higher ionization potential will exhibit larger variations with the T_e. It follows that a rise in electron temperature causes a relative intensification of the lines of the component which is difficult to excite. This is why the helium lines in the spectrum of an argon-helium mixture become more intense as the pressure and the diameter of the discharge tube are reduced. A higher current as well as a higher electron concentration should also intensify the lines of a component which is difficult to excite.* The intensity ratio obtained in a gas mixture subjected to a DC discharge is sharply distorted by the separation of the components (see Section 4).

---

*The effect of discharge parameters on the relative intensities of the spectral lines is discussed at length in Section 15.

Chapter II

LIGHT SOURCES*

4. GLOW DISCHARGE

A *glow discharge* (the common form of discharge at low pressures) is a self-sustaining phenomenon, that is, one which persists after the external source of ionization has been withdrawn. A nonself-sustaining discharge may become one if the number of electrons and ions produced in it is larger than or equal to the number of ions leaving the discharge space. It is the development of electron avalanches that is responsible for the creation of ions in a discharge. Thus, the electrons in the discharge gap ionize atoms upon collision with the latter. This in turn releases other electrons, which again ionize other atoms, until an electron avalanche results. The number of electrons drifting toward the anode increases with the distance from the cathode. The potential difference at which a nonself-sustaining discharge becomes self-sustaining is termed the *firing potential*. It has been shown experimentally (Fig. 10) that the firing potential depends on the product pd, where p is pressure and d the distance separating the

*See also general handbooks on gas discharges [21, 22, 105, 107, 171-174].

electrodes in a discharge tube. The curves in the plot show minima, reflecting a twofold effect. On the one hand, an increase in either the pressure or the size of the discharge gap increases the number of ionizing encounters, and hence facilitates the initiation of a discharge. On the other hand, the mean free path of an electron decreases with pressure so that the electron accumulates less energy between collisions. Thus the probability of ionization by collision diminishes, and the initiation of a discharge becomes more difficult. The firing potential of a discharge depends on the type of gas filling the tube and the electrode material. Minute traces of gaseous impurities materially alter the value of the firing potential.

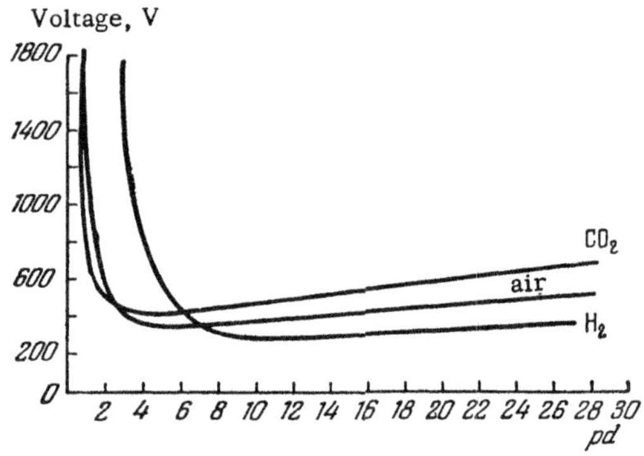

FIG. 10. Firing potential vs. *pd*.

In a glow discharge the cathode remains cold, while a marked cathode potential drop obtains due to the special nature of the distribution of space charges. In glow discharges, lower external resistance produces a higher current density. At some density the potential at either end of the discharge tube begins to drop, the flashover characteristic droops, and the glow discharge becomes an arc discharge, which is characterized by a high current density [174-176].

Under the conditions of a glow discharge the basic processes resulting in a current flow through the gas are ionization by collisions and the release of electrons, some of which are knocked out from the cathode by positive ions while others are ejected as a result of the photoelectric effect associated with the radiation produced by the discharge itself [153, 177].

According to a recent concept advanced by Neu [178], any deficiency in the number of charged particles in a discharge is compensated for by fast atoms, which cause additional ionization. The fast atoms themselves result from charge transfer.

To the observer, a glow discharge appears to be divided into several regions: 1) a thin dark layer, the Aston dark space, in the immediate vicinity of the cathode; 2) a luminous layer called the first cathode glow; 3) the Crookes (or Hiltorf) dark space; 4) the negative glow region, which is extremely bright in the section near the cathode but grows less and less luminous with distance from it; 5) the Faraday dark space; 6) the positive column extending from the anode to the Faraday dark space; and 7) the anode region of the discharge.

The length of the positive column depends on the distance between the electrodes. When the electrodes are moved toward each other, the cathode sections of the discharge remain unchanged, while the positive column is "compressed" and, at a certain interelectrode distance, completely eliminated. When the cathode is displaced in a direction perpendicular to the tube axis, the cathode fall sections of the discharge move with it, but the positive column remains in place, filling all of the remaining space up to the anode. It follows that the charged particles move in an oriented manner in the cathode area and at random in the positive column. As a result, the walls exert an appreciable effect only on the column, but not on the cathode fall.

If the interelectrode distance is shorter than the sum of the cathode sections of the discharge (but all conditions remain as before), then the voltage necessary to initiate the discharge must be considerably higher (i.e., the discharge is difficult to start).

The processes occurring in a glow discharge will be understood better if the reader is acquainted with the details of each of the discharge sections. It is clear from the above that the processes sustaining the discharge occur at the cathode. These are the only processes necessary for sustaining the discharge. Most of the potential drop between the electrodes occurs in the *Crookes dark space*, where the electric field strength is maximum. It is the field at the cathode which causes the electrons to gather momentum and acquire sufficient energy for exciting and ionizing the gas. Ionization produces a large quantity of free electrons and positive ions in the Crookes dark space, but since the velocity of the ions is much lower than that of the electrons, the result is a positive space charge in that section. The dark space ions then impinge on the cathode, knocking out further electrons, the final result of the ionization and ion impact being an electron avalanche. To a first approximation, the size of the Crookes space varies inversely with the pressure; this is because each electron must create enough ions to knock out one electron from the cathode (if that does not happen, the discharge dies out). This means that the size of the Crookes space decreases with pressure: higher pressures produce a greater number of collisions, i.e., more ions per unit length of the dark space.

The *cathode drop*, a basic parameter of the Crookes dark space, is the potential difference between the boundaries of this space. The cathode gradient depends on both the surface ionization and volume ionization coefficients,* and is considerably reduced by the presence of impurities in the gas.

The cathode drop is independent of the pressure or the current up to the point where the entire cathode surface becomes covered by the glow. Beyond this point it increases with the current and current density (*abnormal cathode drop*). It follows that Ohm's law does not apply to a glow discharge. A drop in potential difference with increasing current (a falling current-voltage characteristic) has in fact been recorded for currents large enough to heat up the cathode.

An abnormal cathode drop is associated with *cathode sputtering*, a phenomenon present, although to a lesser degree, in the case of a normal cathode fall. The sputtering is strongly affected by both the electrode material and the gas filling the discharge tube. Sputtering is more pronounced in heavy gases than in light ones, and is stronger for metals of low chemical activity than for highly active ones. In addition, the degree of sputtering is an inverse function of the heat of sublimation of the metal. Detailed studies of this process have shown that it is intensified at lower pressures and larger currents. Cathode sputtering is always accompanied by gas adsorption in the atomized particles, whereby inert gases are adsorbed less intensively than are the chemically active ones. The gas adsorbed by both the vaporized metal and the electrodes themselves can be partially liberated by heating.

The gas adsorption processes in an electrodeless discharge differ from those in a discharge between internal electrodes [179]. In the first case all the adsorbed gas can be liberated by heating the walls to 300°C. In the second case the adsorption never leads

---

*The surface ionization coefficient indicates the number of electrons ejected from the cathode due to the impact of a positive ion. The volume ionization coefficient indicates the number of electrons and ions produced by a single electron as it travels a distance of 1 cm on the path from the anode to the cathode.

to saturation, even when inert gases are involved. Hundreds of monomolecular layers are adsorbed but not liberated on heating. It was shown [175] that gas adsorption is determined by the rate of evaporation of the metal and the surface potential at which metal deposits.

In recent experiments, gas adsorption in a discharge has been studied by radioisotope techniques [180]. Adsorption tests using Kr^{85} disclosed no radioactivity at the anode, but the cathode proved to be radioactive even at a considerable depth (several thousand monoatomic layers). The experiments were carried out in a neon-argon mixture containing 0.0001% Kr^{85}. The discharge tube was operated for 24 hours, the discharge current and potential being maintained at 30 mA and 150 V.

The interaction of gases with the glass and electrodes is described in various communications [181-184].

Cathode sputtering has often proved a source of errors in the spectral analysis of gas mixtures, since the rate of gas adsorption by the metal film formed on sputtering varies for different gases. As a consequence, the composition of the gas mixture may change in the course of the discharge, inevitably affecting the spectral analysis data. Because the gases may be adsorbed and liberated by the electrodes and the sputtered material, the experimenter should forgo, wherever possible, the use of internal electrodes in spectral analysis. Alternatively, he may work with smaller currents and higher pressures in order to reduce the cathode-sputtering effect. The effect of gas adsorption will likewise be reduced if the test is concluded in a stream of gas flowing through the discharge gap.

Adjacent to the Crookes dark space is a luminous region (*cathode glow*) characterized by a very slight potential gradient. In the sector facing the anode the gradient may even be negative;

i.e., the potential may decrease toward the anode. Two factors appear to be involved in this effect. First, due to intensive ionization, the anode end of the glow region contains a large number of free, relatively slow electrons. Second, some of the ionizing electrons in the negative glow region retain a considerable fraction of their energy. The combined action of both groups of electrons can produce the negative gradient. As the distance from the cathode increases, the electron energy diminishes and the glow becomes less bright. Spectroscopic studies have shown that the lines corresponding to lower excitation energies are located in that region of the cathode glow which is closer to the cathode [148]. That region contains many spark lines and, when a mixture of gases is present in the tube, the excited lines are those corresponding to the gas with the larger ionization and excitation energies. This is explained by the presence of electrons which have accumulated high energies (of the order of tens of electron volts) in passage through the cathode fall. The electron concentration in the cathode glow region is much higher than in the other sections of the discharge.

The *Faraday dark space* is located next to the cathode glow region. It emits a rather faint light because the potential gradient in this sector is low and the electrons are thus unable to accumulate energy to replace that lost on passage through the glow region. The electron concentration in the Faraday space, especially in the section nearer the cathode, is as high as in the positive column.

In the vicinity of the anode there is another dark space, which merges into the anode glow region. The *anode potential drop*, amounting to several tens of volts, is not a fixed component of the discharge. It may vanish when the electrons impinge on the anode, that is, when the anode is located inside the glow region produced by the cathode.

The *positive column*, which extends through the entire space between the anode and the cathode fall regions, is the most important light source in the glow discharge. The basic function of the positive column is transmission of the current through the gas. The potential gradient over the length of the column becomes stabilized at a fixed level and, for a constant current density, may be taken to increase inversely with the tube diameter. In a narrow tube the positive ions and electrons reach the walls in less time than they do in a wide tube. The result is a higher recombination rate, which must be compensated for by increasing the number of excitation events which in turn involves a higher longitudinal field gradient.

In a narrow tube a major influence is the transverse field generated by negative charges on the walls, which are stored up as the wall is bombarded by electrons drifting toward it. The cross-field effect produces a curvature in the equipotential surfaces; the latter become convex, protruding toward the cathode. A negative charge on the walls causes ions to drift toward them [143, 185].

At high pressures and large currents, a *pinching (contraction) of the discharge* is noted in the tube: the discharge is compressed into a fairly narrow luminous "rod" along the tube axis. The effect stems from the difference in the gas density between the wall and axial areas. Because of the high heat transfer rate from the walls to the external air, the gas layer adjacent to the walls has a lower temperature, and therefore a higher density, than the gas in the axial region. The lower gas density along the axis, in turn, produces a longer mean free path for the electrons, which facilitates the operation of the discharge, reducing the longitudinal potential gradient in the tube. After its initial establishment, the discharge then narrows further, because the gas core along the axis heats up progressively, and thus the gas density drops off further. The theory of the pinch effect is far from complete at the present time.

Pinching of the discharge may also be caused by a longitudinal magnetic field [186-188]. In that case, the discharge appears to be "flowing" through a narrow "capillary": the latter has no solid walls upon which ion recombination could take place. In many cases a longitudinal magnetic field is a more convenient means of compressing the discharge than is the use of fine capillaries. The effect of such a field on the electron temperature and concentration is negligible [189]. A transverse magnetic field, on the other hand, conspicuously increases the brightness of the discharge.

Compression of the discharge promotes excitation of the ion lines. While all lines are intensified, the degree of intensification varies from level to level, being unequal even for lines emitted by adjacent levels [190]. If impurities are present, the positive column may become striated, with alternating light and dark strata. No definitive theory has thus far been offered to account for discharge striation, although a few attempts have been made [191-193].

Electrophoresis

Luminescence of the gas mixture in a DC glow discharge is associated with *electrophoresis*, an effect which shows up in the increased concentration of the readily ionized constituent (as well as of the component with greater atomic weight) at the cathode. Electrophoretic phenomena have been explored in numerous experimental studies [103, 104, 194-205].

Two theories have been proposed to explain the separation of a binary gas mixture [204]. One of these (the "ion" theory) attributes the separation to a positive ion transport toward the cathode. The stream consists mainly of the component with the lower ionization potential since the number of ions of the other component in the gap discharge is very small [108]. The net result is a greater concentration of the easily ionized constituent at the cathode.

The other ("momentum") theory is based on the assumption that a certain momentum, directed toward the anode, is imparted to the atoms of the gas by bombarding electrons. The transferred momentum is an inverse function of the molecular weight of the gas; consequently, the lighter gas will tend to accumulate at the anode.

The experimental studies of Matveyeva [203] suggest that the separation is due to the difference of ionization potentials rather than in the difference of atomic weights. This is confirmed by the observation that no separation occurs if the concentration of the difficult-to-ionize component in the mixture is low. The fact that the separation depends on the concentration of the readily ionizable component cannot be explained by the momentum theory, but comes out easily from the ion transport theory. The degree of separation increases with the length of the tube and the current. Up to a point, it also increases with pressure and time allowed for separation; however, saturation is reached at a certain time and a certain pressure. This occurs when the gas transport due to ion drift becomes compensated by back-diffusion resulting from the concentration gradient. Thus it takes some time before a concentration equilibrium can become established at the electrodes. Electrophoresis may be used in spectral analysis; it can serve to enrich the gas mixture and to free it of traces of impurities [198, 202].

The gas in the positive column is a quasi-neutral plasma; i.e., the concentration of electrons equals that of positive ions at any given point of the space. The ion concentration (i.e., the degree of ionization) need not be very high. The temperature corresponding to the average electron velocity is several tens of thousands of degrees. On the other hand, the gas itself is at a much lower temperature because the rate of energy exchange between the electrons and the atoms is low. In a glow discharge the high electron temperature produces excitation of lines that cannot be excited in an arc or a spark at atmospheric pressure. The metastable atoms are of considerable importance in glow discharges (see Section 1).

The electron velocity distribution in the positive column generally follows Maxwell's law (see Section 2), although marked deviations from that pattern are occasionally noted under certain conditions.

Whereas the excited lines seen in the spectra of the positive column at moderate current densities are predominantly those of atoms, the glow spectra do display some ion lines as well. This is due to the fact that some faster electrons are present in the glow discharge [206], as well as to the fact that the number of encounters between the electrons and ions is larger than elsewhere because of the high ion concentrations.

The hollow cathode

A special form of the glow discharge is that proceeding in a *hollow cathode* (Fig. 11), and is widely used in spectroscopy. Its application to spectroscopy is due to Paschen [207]. In this arrangement, at certain pressures and currents the glow is confined within the cathode, whereby the entire cathode potential drop takes place in a very thin layer forming at the internal walls of the cathode tube. This permits high electron velocities even at relatively high pressures because the electrons bounce back and forth between the electrode walls, and are able to produce an intense luminescence in the gases. Both ionic and atomic lines are excited inside a hollow cathode. With gaseous mixtures, the use of a hollow cathode results in simultaneous excitation of atomic lines whose excitation potentials are very much different. This ability to excite varying levels simultaneously is the basis for the use of hollow-cathode discharge tubes in spectral analysis.

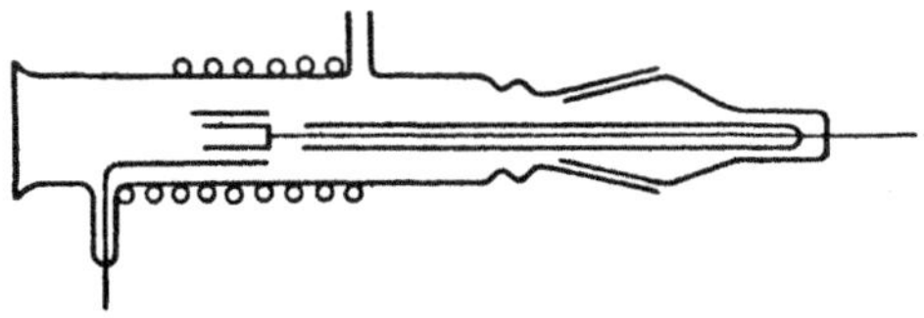

FIG. 11. Paschen's tube with a hollow cathode.

A detailed description of various designs of discharge tubes with hollow cathodes is given in [208–210]. The atomic glow inside a hollow cathode was studied spectroscopically by Frisch et al. [146, 147]. Some data on electron velocity distributions were reported by Veith [211]. A number of experimental studies on the mechanism of discharge in a hollow cathode provided a basis for a tentative discharge theory [153, 155, 212, 213]. It was shown [213] that the specific nature of a discharge in a hollow cathode is related to the high ion concentration and the high velocities of the ions striking the cathode walls. The hollow-cathode discharge closely resembles an abnormal glow discharge.

5. HIGH-FREQUENCY DISCHARGE

The discharge induced in the gap by a rapidly alternating electromagnetic field is called a *high-frequency* (or radiofrequency) discharge. Under these conditions, the charged particles are unable to follow the rapid changes in the field direction. This results in a phase difference between the electric field and particle velocity, a difference proportional to the mass of the moving particles. For practical purposes, the ions existing in a stable high-frequency discharge may be treated as motionless. On the other hand, the motion of the electrons in an AC field, is determined by the ratio of the electromagnetic field frequency ω to the frequency of collision ν_{coll} which the moving electrons experience [172].

For moderate current densities, the effect of the magnetic field may be considered negligible. Moreover, if $\omega \gg \nu_{coll}$ the encounters of electrons with gas particles during a single field cycle may likewise be ignored. An oscillatory motion caused by the field is thus superposed on the random thermal motion of the electrons. The magnitude and direction of the translational component of velocity depend on the phase of the AC field at the moment when the electron begins to move. If the field phase is $\frac{\pi}{2}$ or $\frac{3\pi}{2}$, the field causes the electron to perform harmonic oscillations about a certain equilibrium position. The translational component of the velocity is maximum when the field phase equals 0 or π. The particular case where $\omega \gg \nu_{coll}$ is achieved only with high field frequency and low gas pressure. In such cases very little energy is required to maintain the discharge.

In the case where $\omega \ll \nu_{coll}$, which takes place at high gas pressures and low field frequencies, electron drift sets in, due to the large number of collisions between the electrons and the gas particles taking place during a single field cycle. The conditions of the discharge approximate those which establish themselves in a DC discharge. The quantity of energy required for sustaining the discharge is very high.

In most instances, however, the actual conditions determining the motion of the electrons do not correspond to either of the two extreme cases. As in the first case, an oscillatory motion becomes superposed on the translational motion of the electron, but the effect of impacts of electrons on the gas particles during one field cycle can no longer be neglected. It is also not possible to ignore the changes occurring in the electric field, or the velocity distribution in the moving electrons in the time interval between two successive collisions.

The theory of high-frequency discharges was developed by Holstein [214], Margenau [215] and Hartman [216] on the basis of Boltzmann's gas-kinetic equation. We shall consider here only high-frequency discharges at reduced pressure generated by continuous oscillation since it is this type of discharge that is used in spectroscopy and spectral analysis.

The range of frequencies used for spectrum excitation is extremely broad. Most studies, both early and recent, have used frequencies in the range of several kilocycles to a few hundred megacycles. Only in the past decade and a half have bands in the range 2000 Mc/s to 50,000 Mc/s become popular.

In establishing the theory of the high-frequency discharge it is important to know whether the high-frequency discharge possesses some specific characteristics that distinguish it from other types of discharges and whether the frequency is a discharge parameter equal in importance to the current and the pressure. If the discharge technique is to be applied effectively in spectral analysis, it is important to know the effect of the frequency on the electric parameters of the discharge. It is even more important to know how the variation of frequency affects the excitation of components of a mixture.

The type of coupling between the oscillator circuit and the discharge gap depends on the exciting frequency range. At frequencies below 2500 Mc/s, the power is supplied via external or internal electrodes (except in the case of an electrodeless ring discharge). When working with frequencies above 2500 Mc/s, the discharge tube is installed in a waveguide. The type of coupling determines the distinctive characteristics of a high-frequency discharge, since it determines the amount of power that can be fed to the discharge gap. Conventionally, two types of high-frequency discharge are distinguished at low pressures: glow discharge and ring discharge [22, 172].

High-frequency glow discharge

This is generated in a gap connected to a high-frequency circuit through internal or external electrodes. The external observer will see several clearly defined regions of the discharge. The positive column is located in the section midway between the two electrodes, both of which act as cathodes, each carrying the full complement of regions of the cathode fall. The oscillatory motion of the electrons in the discharge tube causes intensive ionization, eliminating the necessity of electron ejection from the cathode; i.e., the electrode processes are of no vital significance. The regions of a high-frequency discharge located near the electrodes are identical with the cathode fall regions of a DC glow discharge [217, 218]. Lodge and Stewart [218] studied the distribution of deposits on the discharge tube walls for the case of a high frequency discharge generated through external electrodes. The experiments revealed the development of a process analogous to the cathode sputtering present in a DC glow discharge.*

In discharges using external electrodes, the function of the cathode is performed by a section of the discharge tube located inside the outer electrode. The strong radial field inside the

*The mechanism of sputtering of internal electrodes in high frequency discharge was studied by Levitskiy [219], using the range of 1-70 Mc/s and atmospheres of argon and hydrogen.

electrode causes an ionic bombardment of the "glass cathode" [220]. In consequence, a kind of "hollow cathode" is formed inside the electrode. The assumption seems all the more plausible if we consider the differences in spectra obtained inside the electrode space and in its vicinity [221]. Thus, inside the electrode one observes luminescence of those nitrogen bands that require very high energies for excitation. Also, it is possible to excite the He II spark line of λ = 4686 Å (excitation potential 75.6 eV) inside a high-frequency "hollow cathode." The above shows that basically there is not much difference between the high-frequency and the usual type of glow discharge.

The parameters of the positive column of high-frequency and DC glow discharges were compared in a series of studies. The firing potential of the high-frequency discharge is considerably smaller than that of the DC system, and varies with the frequency [222, 223]. The effect of the walls on the firing potential seems especially important [224, 225].

Beck [226] compared a high-frequency discharge (wavelength 5 m) with a DC discharge (potential 1200 V), using a discharge tube with a hot cathode and a nickel anode. A 50-mA current was employed in both cases. In both cases, Hg lines (λ = 5461 Å) of equal intensity were obtained. The ratio of intensities of all lines of the mercury spectrum and the Hg line with λ = 5461 Å proved to be identical at three different pressures (10^{-3}, $7 \cdot 10^{-3}$ and 0.3 mm Hg). The electron temperature values obtained by Beck via probe measurements for high-frequency and DC discharges were identical at high pressures, and only at low pressures was a slight divergence noted.

The probe measurement data reported by Beck are consistent with those obtained in later studies [87, 227, 228] in which the electric parameters (T_e and n_e) of high-frequency and DC discharges were compared. Thus, the comparison of DC and high-frequency (5-Mc/s) discharges in helium, neon and argon was made by Dzerpetov and Pateyuk [227], while Avramov and Dzerpetov [87] studied it in helium and neon. The values of T_e and n_e obtained in the two types of discharges were identical. Similar results were obtained when working with a high-frequency (170-Mc/s) discharge at pressures in the range 0.26-1 mm Hg [228].

The effect of the frequency (over the range of 0.7-70 Mc/s) on the electron concentration for a high-frequency discharge in mercury was studied by Levitskiy and Shashurin [86]. When the frequency was varied from 0.7 to 20 Mc/s the value of n_e was reduced by a factor of 4; on further increase to 70 Mc/s, this value increased by a factor of 2.5.

Bhatt [229] and Nöller [230] noted that the spectra in high-frequency discharges do not differ from those obtained at other frequencies. In Nöller's experiments, the effect of a high-frequency discharge in helium, observed over the medium-wavelength range ($\lambda = 75$ m), was compared with that of a short-wave discharge ($\lambda = 30$ cm). The author concluded that the ratio of intensities of spectral lines does not vary with the frequency. However, Nöller's conclusion is not well documented since the lines he selected for comparison are not particularly sensitive to variations in discharge conditions.

We ourselves tested the effect of variation of the exciting frequency on the ratio of spectral line intensities in gas mixtures whose constituents differed sharply in their ionization potentials. We used a mixture of 5% argon in helium in a 2-mm I.D. capillary at a pressure of 0.5 mm Hg, varying the wavelength from 9-92 m. We saw no evidence of any changes in spectra which could be associated with the frequency variation, in contrast to the results reported in many other communications [231-234]. However, in varying the frequency it is quite difficult to ensure constancy of all the other discharge parameters. We cannot be certain, therefore, that the differences in spectra observed by the other authors were directly related to the frequency of the exciting system.

The differences in the behavior of electrical parameters of high-frequency and DC discharges become apparent only at high frequencies and relatively low gas pressures.

It would thus appear that in cases where $\omega \ll \nu_{coll}$ the high-frequency discharge does not differ from the DC discharge, and no effect of frequency on the spectra is noted over an extremely wide range of frequencies and pressures. Nonetheless, there is a clear advantage in using a high-frequency discharge for spectral analysis. It eliminates the need for metal electrodes inside the discharge tube, thus reducing gas adsorption during the discharge and eliminating the possibility of a chemical reaction with the electrode metals. Moreover, a high-frequency discharge can be started at very low pressures where the excitation of gases with high ionization potentials is most favorable.

It was shown [235-237] that there exist two types of high-frequency discharge ("weak" and "strong") that differ in their optical and electric characteristics. The transition from one type of discharge to the other can be discontinuous. The pressure range in which both kinds of discharge can exist varies with the nature of the gas and the diameter of the discharge tube. In a "weak" discharge it is the lines with the higher excitation potentials which are more intense; the opposite is true for "strong" discharges. Also, the concentration of excited atoms is also higher in the case of strong discharges. The observed optical differences are due to the difference in electron concentrations; however, the electron temperatures appear to be identical. The electrical parameters of weak and strong discharges were compared in a special study [236]. The existence of two kinds of discharge was verified by Levitskiy [223], who tested the effect of pressure on the firing potential of a high-frequency discharge in argon and hydrogen, using the transitional frequency range. The curves of pressure vs. firing potential displayed an abrupt rise separating the two types of discharge. Both forms may also occur when a longitudinal magnetic field is superposed on a high-frequency discharge [223, 235, 236]. In this case, a weak discharge may transform to a strong one, with the latter then sustained regardless of whether the magnetic field is present or not.

In recent years the superhigh-frequency (SHF) discharge has been steadily gaining in popularity [238, 239]. This range is generally understood to include frequencies between 10,000 and 50,000 Mc/s * (the millimeter, or ultrashort wave, range). In the superfrequency band (more exactly, at its lower boundary) the oscillation amplitude of the electrons becomes smaller than the dimensions of the discharge tube. This determines the two distinctive features of a superfrequency discharge: 1) the discharge is carried exclusively by electrons; and 2) electrode and the tube wall processes are inconsequential.

Vashman et al. [241] noted significant changes in the discharge conditions upon transition from 6 Mc/s to 3000 Mc/s. In a helium-argon mixture, the employment of superhigh frequencies caused a better than twofold increase in the ratio of intensity of the 5876-Å helium line to various argon lines. This imples that the sensitivity of spectral analysis for the difficult-to-excite component of the mixture should increase when the decimeter range is employed.

---

*The special features of the superfrequency discharge are occasionally observed even at lower frequencies, that is, the centimeter range [240].

The superfrequency discharge conditions differ drastically from those in other types of discharge [238]. A striking characteristic is the extremely high electric-field component [242]. An increase in the general brightness of the discharge with frequency was observed in numerous experiments [240, 243, 244]. The effect of the oscillator frequency on gas adsorption was studied by Meggers and Westfall [240]. Thus, an increase in frequency from 30 Mc/s to 90 Mc/s resulted in a nearly three hundredfold decrease in adsorption of gases by mercury, and in the frequency region above 100 Mc/s, no perceptible gas adsorption was recorded.

Extensive spectroscopic studies of the high-frequency glow discharges have been made [99, 126, 245-247]. Cabannes [246] demonstrated that in a large-diameter discharge tube, the spark lines are excited only near the axis—an indication of a nonuniform column cross section. Harrison [126] observed a strong reabsorption in hydrogen lines even when the hydrogen constituted only 1% of the mixture. Studies of the high-frequency discharge in neon [99] disclosed that reabsorption may involve even the lines terminating at the $2p^5 3p$ levels. All the observed facts point to a high concentration of excited atoms in high-frequency discharges.

Ring discharge

A high-frequency discharge may be produced in the form of a ring generated in a discharge tube placed directly in an alternating magnetic field. The process is similar to induction of a current in a solid metal. The theory of the electrodeless discharge was worked out by Thomson [248] and Mierdel [249]. A schematic diagram of Thomson's circuit for the excitation of a ring discharge is shown in Fig. 12. This basic arrangement has been used by all later experimenters with only minor modifications. While the first spectroscopic study of glow discharge luminescence dates back to 1913 [250], it was Thomson who described the essential characteristics of the ring discharge. Thus, the ring discharge can exist only in an extremely narrow range of pressures. The effects of frequency and pressure on the conditions required for producing a ring discharge in various gases were studied by Strauss [251]. Experiments on starting a discharge in a toroid were conducted by Davydov [252], while Birkkhoff [253] investigated the electrical processes involved in a high-frequency ring discharge.

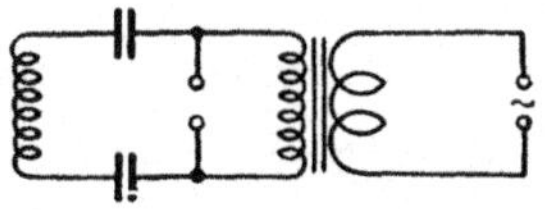

FIG. 12. Thomson's spark generator.

The power consumption in this discharge is determined by the length and capacitance of the discharge gap. In general (although there are some exceptions) [254], the distribution of excitation energies through the cross-section of the ring is nonuniform, with spectral lines of higher excitation energy being concentrated in the outer layers while those of lower energies tend to stay in the inner layers. The ring discharge has thus proved useful in classification of spectra [255-258], since it allows the simultaneous observation of lines produced by both normal and ionized atoms. If the image of a luminous ring is projected onto the slit of a stigmatic spectrograph then the spectrogram will show both long (arc) and short (spark) lines. As the length of the discharge gap is increased the spark lines, produced by single-charge ions, become longer and, in addition, short lines representing doubly-ionized atoms are observed. The possibility of simultaneously observing spectral lines with different excitation energies suggests that the ring discharge may prove extremely valuable in spectral analysis, especially analysis of gases, where the basic difficulty has been to produce simultaneous luminescence of two components of a mixture which differ sharply in their ionization potentials.

However, a spark generator based on the Thomson circuit can hardly be expected to maintain constant discharge conditions, especially since long discharge gaps (7–10 mm) are required in order to obtain large excitation energies. The sine wave vacuum tube oscillator seems preferable in this respect. However, its power must be very high. Thus pulse oscillators are preferred [51, 259] because these are capable of generating pulses of very high power while the average power of the output is rather moderate. In addition, the vacuum tube oscillators are capable of maintaining constant conditions in the discharge.

In one arrangement [260], pulses are generated at low voltage and high capacitance. An electrodeless tube filled with gases or vapors at a pressure of a few hundreths of a millimeter is placed in the center of a coil through which pass high-frequency currents. At pulses with power lower than 10 kW only the arc lines are excited. The average discharge power is about 500 kW. Spark lines appear only when the pulse voltage is increased. On the negative side, when a pulse oscillator is used to generate high-frequency discharges (either the ring or glow type), the field strength required for starting the discharge must be very high, the strength varying inversely with the pulse duration τ.

6. PULSE DISCHARGES IN TUBES WITH INTERNAL ELECTRODES

The pulse discharge is characterized by high current densities and a high degree of ionization. Thus, when a pulse discharge is produced by discharging a condenser (fed by a source of DC current) through the gap, one can obtain spectra of doubly and triply ionized atoms, even though the average power of the discharge is low. The circuit diagram of a pulse generator is shown in Fig. 13. It is seen from the figure that the condenser is charged until a certain potential difference is reached and then discharges through the discharge gap. Circuits of this type have been used by many experimenters [261-272]. To increase the limiting potential to which the condenser can be charged, a sparkover gap is connected in series with the discharge tube. Thus, the potential across the tube can be higher than the firing potential. Instead of the spark gap, the firing can be controlled by means of a simple electronic circuit, shown in Fig. 14 [265]. The pulse duration may thus be varied from 10^{-6} to 10^{-3} seconds. The discharge power is determined by the condenser capacity and the potential difference at the condenser at discharge. The electric properties of the pulse discharge are discussed in detail by Laporte [272].

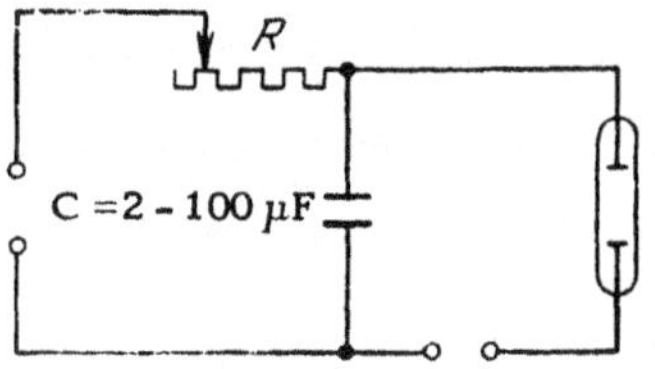

FIG. 13. Pulse oscillator with a spark gap.

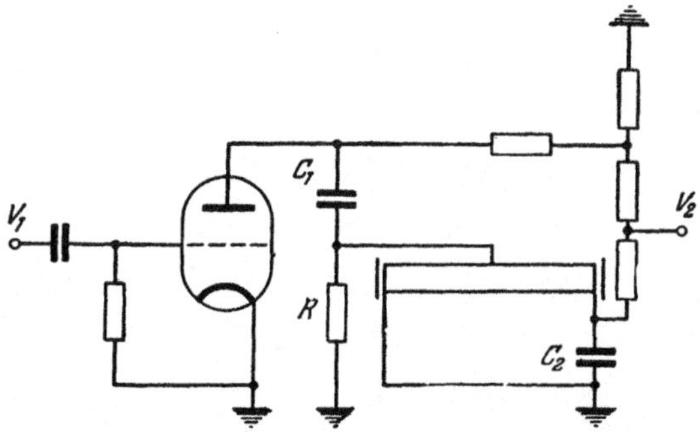

FIG. 14. Circuit for control of pulse discharge. $C_1 = 0.01$ μF; $C_2 = 0.56$ μF; R = 100 ohms; V_1 = 150–200 V; V_2 = 25–50 kV.

The pulse discharge mode can be either oscillatory or aperiodic, depending on the capacitance, self-inductance and resistance of the circuit. The conditions under which the oscillatory mode changes to aperiodic are given by the general relationship for a conventional oscillatory circuit: if $C > \frac{4L}{R^2}$ the discharge becomes aperiodic.

It follows that an increase of the circuit capacitance, at a given self-inductance and resistance, may result in transition from the oscillatory to the aperiodic mode.

The value of the peak current rises with increasing capacitance until saturation is reached. A further increase in capacitance will not produce a higher peak current but will increase both the pulse duration and its energy. The maximum current density, even when the average density is relatively low, may be as high as several thousand amperes per square centimeter. To obtain large current maxima it is necessary to reduce both the inductance and the effective resistance of the circuit. Pulse condensers with a low self-inductance are employed most advantageously in this case. Tubes producing very bright discharges have constant resistances amounting to several ohms [264, 268].

At low pressures the brightness of the discharge is considerably lower. Intense pulses of short duration are obtained at low capacitances and high voltages. The composition and the energy of the discharge radiation depend on the nature and pressure of the gas, the tube diameter, and the electric parameters of the discharge. The radiant background, for example, is less intense in narrow than in large-diameter tubes, and its intensity is higher in tubes filled with a heavy rather than a light gas. There is practically no background when the discharge tube is filled with helium.

The pulse discharge parameters were studied in inert gases by Vul'fson and Bogdanov [263, 264]. Their circuit consisted of a condenser connected in series with a thyratron and a discharge tube. They have shown that the pulse discharge, in contrast to the arc and spark, is characterized by a very large potential drop and therefore a large potential gradient in the positive column (which may reach several hundred, or even thousand, volts per centimeter). The high-power discharge results in complete ionization of the gas. Only a few atoms are ionized immediately after the breakdown, when the energy transmitted to the gas is still small. However, the number of ionized atoms grows as the energy stored in the condenser is dissipated in the discharge gap. Intense ionization is accompanied by a drop in electron temperature and a simultaneous rise in the atomic and ionic temperatures. By the end of the pulse the electron and atom temperatures are equal. Following the discharge, a recombination process sets in, and energy is dissipated by convection losses and because of the heat conduction by the gas.

A pulse discharge, unlike an arc, excites some ionic lines along with the atomic lines. Studies of pulse discharges by the time-resolved spectrum method [263] showed that spark lines are excited earlier than arc lines. This method is described in detail in the literature [273, 275]. Figure 15 shows a time-resolved spectrum of radiation from a krypton tube (length 50 cm; diameter 10 mm; pressure p = 5 mm Hg) [263].

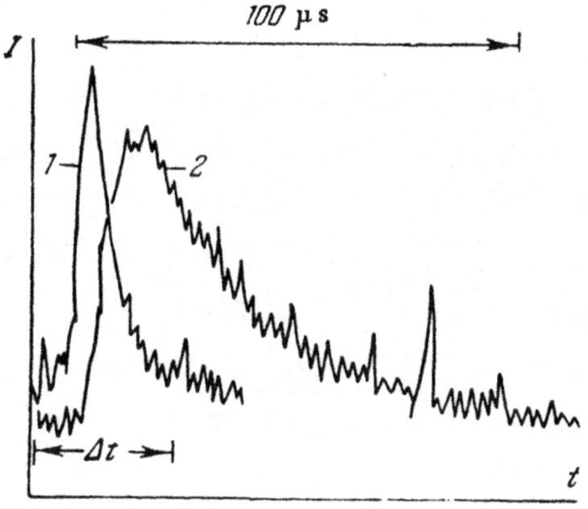

FIG. 15. Time-resolved radiation spectrum from a krypton tube. 1—Kr II, 4292.94 Å; 2—Kr I, 4273.96 Å.

So rapid is the "heating" of the gas that the curves in the time resolution of the spectrum record only its "cooling."* If the "heating" proceeded more slowly, a plasma warmup period would also be recorded, and the arc lines would reach maximum intensity before the spark lines, since they require less excitation energy. As seen from Fig. 15, the spark spectrum is the first to be excited. Its peak intensity lags the current maximum, and is reached when most of the power has been dissipated in the discharge gap. The arc spectrum, on the other hand, does not appear until recombination has set in. This occurs only after the end of current passage; therefore, the intensity peak of an arc line is shifted to the right of the spark line peak.

High-power pulses can also be obtained in a tube in which the breakdown of the discharge gap is achieved by means of a third electrode [261, 269]. The charged condenser is connected to the discharge tube electrodes, but the firing potential of the discharge is much higher than the potential existing across the condenser. Hence the breakdown does not occur until a potential is established at the third electrode by means of a special transformer. It is also possible to eliminate the third electrode by placing the tube inside the inductance coil of a high-frequency circuit [270]. An alternative, simpler method is to connect a wire touching the glass tube to a Tesla transformer [271].

Thus the pulse discharge studies show that we are dealing here with a highly ionized plasma. The instantaneous discharge power reaches enormously high levels. This creates conditions which facilitate the excitation even of gases with very high ionization potentials.

---

*The time-resolved spark spectra show an analogous behavior [275].

Chapter III

SPECTROSCOPIC EQUIPMENT. PREPARATION OF STANDARD MIXTURES

7. THE VACUUM UNIT

Spectral analysis of gases generally requires the availability of a vacuum unit. The latter, besides allowing the introduction of the sample into the discharge tube, serves a variety of purposes such as evacuation of the discharge tube and of the entire system, preparation of standard mixtures, pressure control, etc.

A vacuum unit (see Fig. 16) is best suspended on a metal rack. The auxiliary equipment, that is, cylinders with pure gases or standard mixtures, is also suspended. After having been welded together, the unit must be thoroughly degassed by heating all connector tubes with the smoky flame of a gas burner. Quartz discharge tubes can be brought to red heat. Benzene, water and stopcock grease vapors interfere with the analysis, and especially with the hydrogen determination since they decompose in the discharge, liberating hydrogen. Therefore, the greasing of stopcocks must be followed by prolonged evacuation of the system. When stopcocks already in use develop a leak and must be resealed, the best thing is to deposit a new coat of grease over the old one without washing the old grease with ether or benzene. One thus obviates a time-consuming reevacuation of the system.

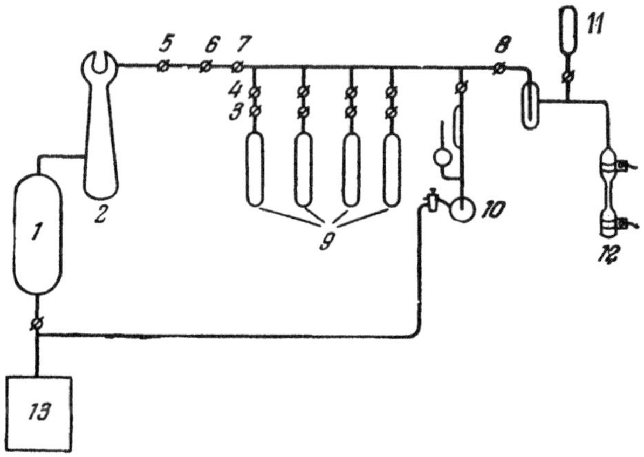

FIG. 16. A vacuum system for spectral analysis of gases. 1—Forechamber; 2—diffusion pump; 3-8—stopcocks; 9—gas cylinders; 10—McLeod gage; 11—auxiliary (surge) vessel; 12—discharge tube; 13—forepump.

The system and discharge tube must be evacuated down to pressures of the order of 10^{-5} mm Hg. Work with extremely low concentrations, as well as microanalysis, requires even lower pressures, which are achieved with rotary oil pumps and metal or glass diffusion pumps (mercury or oil vapor).*

Before an oil vapor pump can be used, the discharge tube must be evacuated directly by means of a rotary pump. During this step atmospheric air enters the tube at the forevacuum pressure. Then the oil vapor pump removes mercury vapor whose presence reduces considerably the sensitivity of the analysis for the difficult-to-excite components of the mixture (the vapor pressure of mercury at room temperature is about $2 \cdot 10^{-3}$ mm Hg, and that of

*See general handbooks on vacuum techniques [276–283].

oil, $5 \cdot 10^{-7}$ mm Hg). When working with a mercury diffusion pump it is often necessary to freeze out the mercury vapor by means of intermediate traps cooled with Dry Ice (−79°C). Liquid nitrogen should not be used because it reduces the temperature to a level at which gases (especially heavy inert gases) are strongly adsorbed by the glass of the system. When this happens, the composition of the sample in the discharge tube changes with time. Freezing out with Dry Ice is adequate because at this temperature mercury has a vapor pressure of 10^{-7} mm Hg. If Dry Ice is undesirable for some reason, traps filled with metallic potassium or sodium can be used if the sample contains only inert gases [277].

A short-type precalibrated [284] McLeod gage with two or three capillaries [277] is used to measure sample pressures. It is useful in the range from 10^{-4} to a few mm Hg. Pressures above 5-6 mm Hg are measured with a U-shaped mercury- or oil-filled manometer. The possibility of introducing errors into the measurement should be considered when using a McLeod gage to measure the pressures of gases (carbon dioxide or steam (near their liquefaction points [285].

The accuracy of pressure measurements with a mercury manometer depends on the accuracy of reading of the mercury level. Various optical arrangements have been devised to facilitate the reading of the capillary scale. Simple interference methods for accurate and rapid reading of the mercury level in the capillary were described by Terrien [286].

Ionization or resistivity (Pirani) vacuum gages are seldom used in gas analysis since they must be recalibrated whenever the composition of the gas mixture is changed. However, a method for measuring the pressures of different gases with the same Pirani gage was suggested by Veis [287].

Moderate and high vacuum measurements can also be made with capacitive diaphragm gages whose readings do not depend on the type of gas used. The theory, design and measurement procedure used with these gages have been described by Drawin [288]. This gage is especially suitable for industrial applications.

The pressure existing in the vacuum system can be roughly estimated from the color of the discharge produced in the connecting tubing or in a spark gap fed by a Tesla transformer. Thus, the air glow changes from reddish-pink to sky-blue as the pressure decreases. The glow ceases at very low pressures ($\sim 10^{-3}$ mm Hg). The Tesla apparatus also allows detection of leaks in a vacuum unit.

Particular attention should be given to the vacuum stopcocks. All of these must be prechecked for leaks, keeping in mind that a stopcock may pass not only the atmospheric air, but also gases from the adjacent compartments of the unit. To test for leaks in, for example, the system of Fig. 16, gas is admitted from cylinders through stopcocks *3* and *4*; stopcock *4* is kept closed while stopcock *3* is open. The gas from the cylinder now fills the space between stopcocks *3* and *4*. Stopcock *3* is then closed and stopcock *4* is opened so that the gas from space *3* and *4* fills the space to stopcock *7*, which must be closed. Batch-metering stopcocks (Fig. 17) are best for measuring the amount of gas. These differ from standard vacuum stopcocks only in the design of the plug. The latter is sealed at the bottom, and the pocket replacing the lateral opening has a capacity which varies from one mm^3 to several cm^3.

Mirtov and Boykov [289] designed an electrically operated stopcock for convenient and precise metering of gases. The device is capable of good performance over a wide range. The stopcock, or valve, is shown schematically in Fig. 18. The basic component of the valve is a capillary with a short porcelain tip *1* fused to it at one end. The capillary connects the two valve chambers which are separated by a partition. The upper chamber connects to the gas cylinder, the lower to the system. A cap *2*, containing an iron core *3*, is located in the upper section of the valve. The upper and lower chambers are filled with pure mercury *4*. Should air at atmospheric pressure penetrate into the system, the mercury in the lower chamber will seal the capillary. The valve operates in the following manner: in its normal position, cap *2* dips into the upper mercury pool which rises to level h_1 and covers the porcelain tip. To let a measured quantity of gas into the system, the solenoid *5* is switched on. This lifts the cap, the mercury level drops to h_2, exposing the porcelain tip. The gas then diffuses through the tip into the capillary and then bubbles through the mercury into the system.*

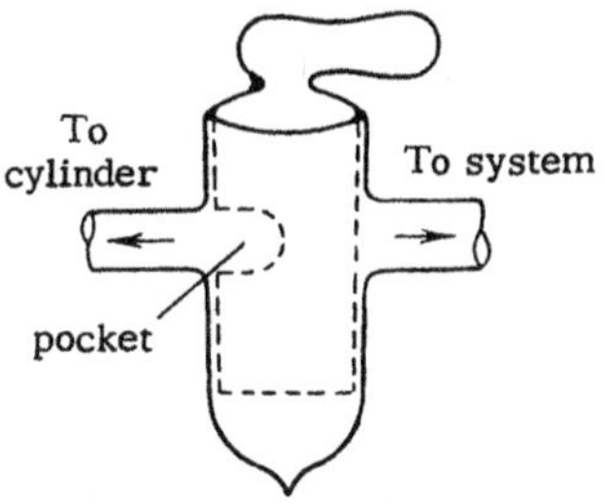

FIG. 17. Batch-metering stopcock with pocket.

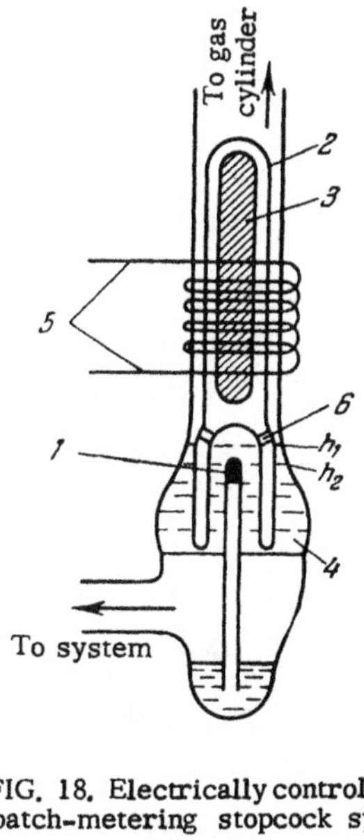

FIG. 18. Electrically controlled batch-metering stopcock system.

The rate of diffusion can be controlled by using porcelains of varying porosity. In a precalibrated valve the gas can easily be metered on the basis of time of flow. The valve requires no lubricant of any kind and the mercury vapor is easily frozen out. However, if the valve is used to meter a multicomponent gas mixture it is essential to check whether the ratio

*Holes *6* in the cap permit the gas to escape from the space underneath the cap when the solenoid is switched on.

of components is preserved (the diffusion rates of the gaseous constituents may differ because of difference in molecular weight).

A simplified vacuum system such as the one shown in Fig. 19 is most suitable for gas analysis under industrial conditions (see Section 26).

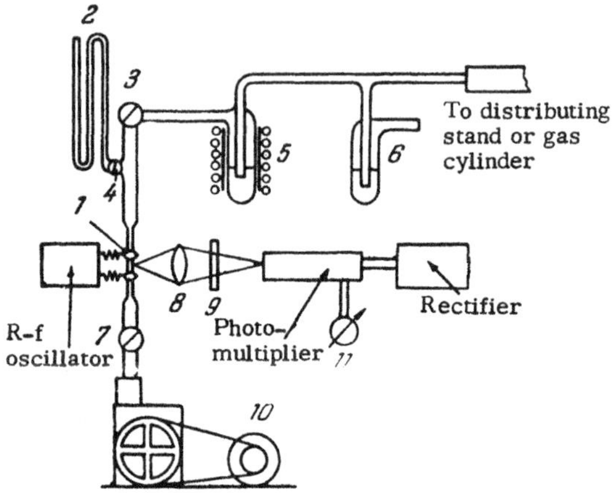

FIG. 19. Simplified vacuum system. 1—Discharge tube capillary; 2—mercury manometer; 3, 4, 7—vacuum-type stopcocks; 5—trap filled with copper shavings; 6—gas vent; 8—collimator; 9—filter; 10—forepump; 11—microammeter.

A continuous gas stream is drawn through discharge capillary *1* by means of a conventional forepump *10*. The gas pressure is measured by means of a U-type mercury manometer *2*. The gas flow rate is adjusted with the aid of vacuum stopcocks *3* and *7*. Stopcock *4* seals off the manometer from the system, if needed. The excess gas is vented to the atmosphere through vent *6*.

Pressure and gas flow rate through the discharge tube can be controlled by a set of parallel capillaries located upstream and downstream from the discharge tube. Such a system is used for determination of isotopes of hydrogen. Capillary internal diameters are chosen on the basis of the desired flow rate and operating pressure. A unit for the calibration of capillaries is described in

[290]. The usual flow rate in the isotopic analysis of hydrogen ranges from 0.1 to 1.0 cm^3/min at capillary bores ranging from 0.005–0.02 cm. The capillaries are bypassed when evacuating the system.

The use of gas-leak valves in place of vacuum stopcocks provides a more convenient means for setting up a continuous gas flow [291]. One valve of such type, a bellows-operated control valve, was described by Astrov and Voronel' [292] and permits regulation of the gas flow over a range of pressures up to 150 atm. Its principal component is a stainless steel stem *1* (Fig. 20) carrying the actual working disc. The latter consists of a conical section which serves to cut off flow, and a 3.5 mm long, 2 mm I.D.

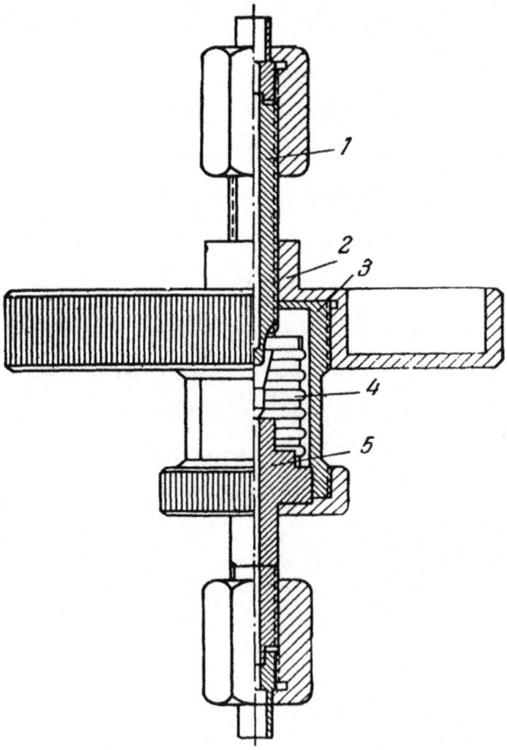

FIG. 20. Sylphon control valve.

cylinder below it. The cylinder tapers to a truncated cone (angle of taper: 60°), and fits with a small clearance into the valve seat *5*. The seat—cylinder combination is the actual control element. The stem is coupled to a stainless steel bellows *4* located in valve body *3*, and is moved by means of handle *2* carrying a double thread (pitch differential—0.25 mm). This valve permits close regulation of the gas flow in the range of 0.03 to 3 cm^3 (STP)/min, at a valve pressure drop of up to 150 atm and a control accuracy of 0.1-0.2 cm^3/min. Other types of valves can also be used to regulate the gas flow.

8. DISCHARGE TUBES

A gas mixture is excited to glow in discharge tubes whose design varies with the specific objective of analysis. Several types of discharge tube used in spectroscopic gas determinations are shown in Figs. 21-25.

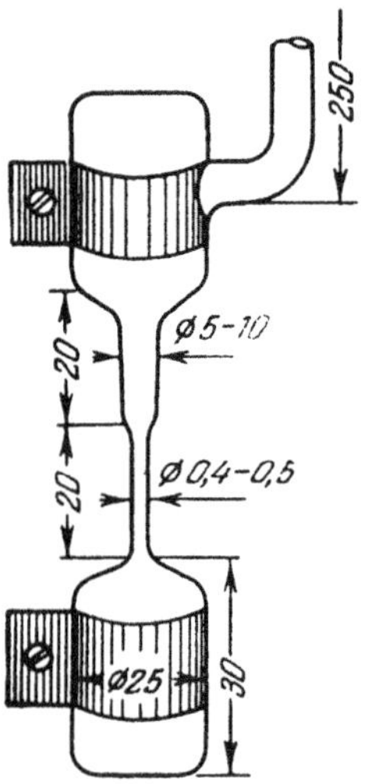

FIG. 21. A fused quartz tube with variable cross section.

When the excitation source is an r-f oscillator, a fused quartz tube with external electrodes (Fig. 21) is generally used. The external electrodes are strips of tinfoil, which may or may not touch the quartz and are connected to the output terminals of the r-f oscillator. The quartz tube is welded to the remainder of the system via a graded quartz-glass seal, or it may be cemented in.* In the latter case the quartz tubing connecting to the system must be at least 25 cm long, to prevent the discharge from touching the cemented junction and producing

*In some cases polyethylene tubing can be used to connect the quartz tube and the glass system.

vapors from the cement which may then drift into the discharge tube. The diameter of the working part of the tube varies from about 10 mm to about 0.5 mm. This arrangement permits analysis of small quantities of components of gas mixtures, with the readily excited gases analyzed in the larger section and the difficult-to-excite components in the narrower section. When working with microquantities of gas, the sealed end of a 0.5 mm capillary about 10 cm long will provide a suitable discharge tube (Fig. 22).

FIG. 22. Quartz capillary for microanalysis of gases.

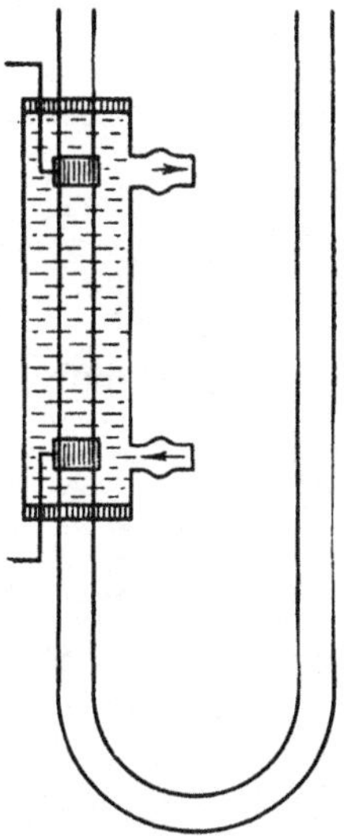

FIG. 23. Discharge tube for isotopic analysis of hydrogen.

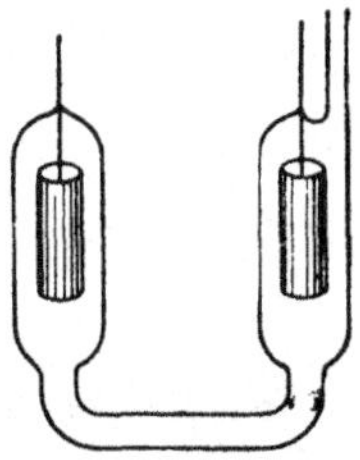

FIG. 24. Discharge tube with internal electrodes (pulse discharge).

The discharge tube shown in Fig. 23 is used for isotopic analysis of hydrogen. The capillary is enclosed in a jacket cooled by running water.

When pulse discharge is used, the tube is made of molybdenum glass (Fig. 24). The narrow section is about 15 mm in diameter. The cylindrical electrodes are made of nickel sheet and are fused to the inside of the tube.

Another type of discharge tube useful for gas mixture analysis is that with a hollow cathode (Fig. 25).

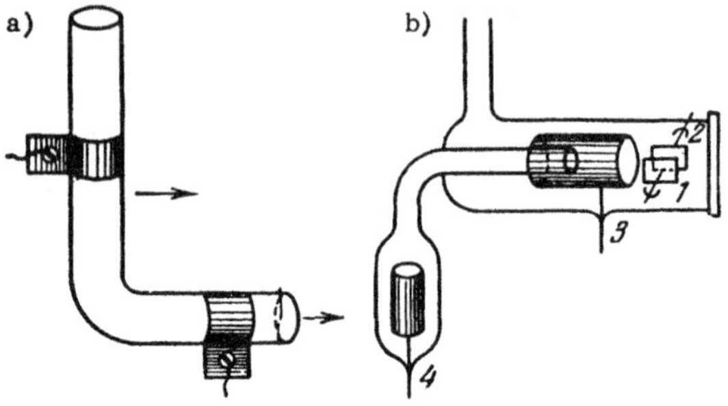

FIG. 25. Hollow cathode tubes: a) high-frequency hollow cathode; b) Frisch's all-purpose tube. (Arrows indicate the direction of photographic exposure.)

In some special cases such as, for example, the analysis of gases for vacuum tubes in electronic industry, the discharge tube may consist of the actual device (i.e., the vacuum tube) [293, 294].

Discharge tubes of diverse designs adapted for various analytical tasks are described in greater detail in Chapter V, which deals with specific techniques of gas mixture analysis.

9. PREPARATION OF STANDARD MIXTURES

Spectroscopic determinations of gas mixtures require the use of standard samples. The standard may be either an already available mixture or a mixture especially prepared from pure components. The minor components of an already available gas mixture are determined by the method of additions (see Section 17). Such mixtures provide particularly convenient standards for rapid gas stream analyses in industry, as well as for analyzing multicomponent gas mixtures.

The simplest vacuum system for preparing standard gas mixtures is shown in Fig. 26. Gas cylinders *A* and *B* are each connected (by means of ground glass joints) to a channel of the vacuum system. Each channel contains a precalibrated vessel of known capacity (V_A and V_B) and a manometer (M_A and M_B). The receiver for the standard mixture ($A + B$), together with the tubes connecting

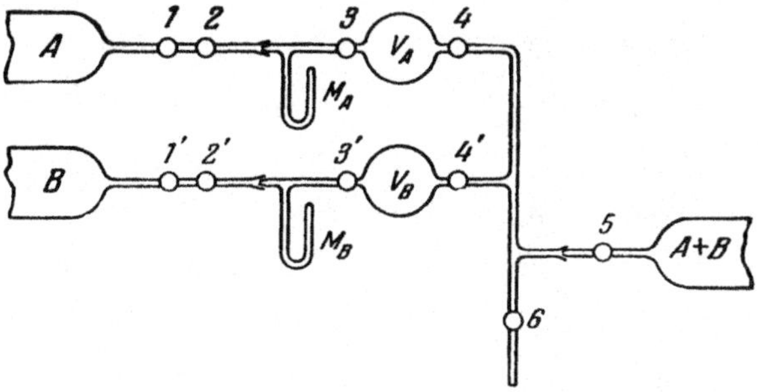

FIG. 26. Installation for the preparation of standard mixtures.

stopcocks *4*, *4'*, *5* and *6*, has a capacity of V. Vessels V_A and V_B are filled, respectively, with gases A and B at pressures p_A and p_B. The remainder of the system is evacuated. The following partial pressures of gases A and B are established when stopcocks *4*, *4'* and *5* are opened (while stopcocks *3*, *3'* and *6* are kept closed):

$$p_{A'} = p_A \frac{V_A}{V + V_A + V_B}, \tag{3.1}$$

$$p_{B'} = p_B \frac{V_B}{V + V_A + V_B}. \tag{3.2}$$

The volumetric concentrations in the mixture are then related by:

$$\frac{C_A}{C_B} = \frac{p_A}{p_B} \frac{V_A}{V_B}. \tag{3.3}$$

Let us now consider a particular case: assume we wish to determine helium in argon, the concentration of the former ranging from 0.1 to 1%. The standards which we shall use should contain approximately 0.1, 0.3 and 1% helium. The equipment needed (apart from the vacuum system and the stand) consists of three empty cylinders of any desired capacity, a cylinder with helium and another with argon, with precalibrated vessels welded to the latter two cylinders. The respective capacities of the two precalibrated vessels are about 1-2 cm^3 (for helium) and 500 cm^3 (for argon). A manometer is connected to each cylinder. For greater accuracy, the pressure should be read off a mirror scale or with the aid of a cathetometer.

Let the helium and argon pressures in the precalibrated measuring vessels be equal. Let it be further assumed that the respective capacities of these vessels are 1 cm^3 and 500 cm^3. If helium and argon contained in the measuring vessels are now transferred to an empty cylinder, the mixture obtained will contain 0.2% helium. By proper selection of pressures in the measuring vessels, the composition of the mixture can be varied as desired. The accuracy achieved in obtaining a given composition of a standard mixture depends on the accuracy with which the capacities of the measuring vessels are known, the accuracy of manometer readings, and the rate at which the gases are mixed.

To determine the capacity of a measuring vessel, the latter is filled with mercury, weighed, and the capacity calculated from the density of mercury at a given temperature. The connecting tubes and stopcock orifices must be at least 5 mm in diameter so that gas mixing will be fast. The standard mixtures are stored in glass containers, which are then sealed onto the vacuum system.

The preparation of standard mixtures is by the Töpler pump [290, 295], shown schematically in Fig. 27. Container *C* (mixture receiver) is mounted on a ground glass joint *2* on top of the Töpler pump, and the mixture components are transferred to it one by one from chamber *F* by means of mercury. Air admitted to chamber *4* through the three-way stopcock *1* drives the mercury upward in the rise of the Töpler pump. When *4* is evacuated with a forepump, the mercury level drops. Pressure measurements

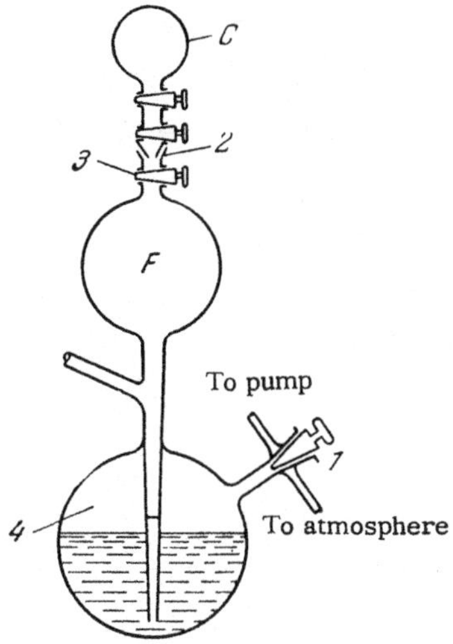

FIG. 27. Mercury compressor for the preparation of standard mixtures.

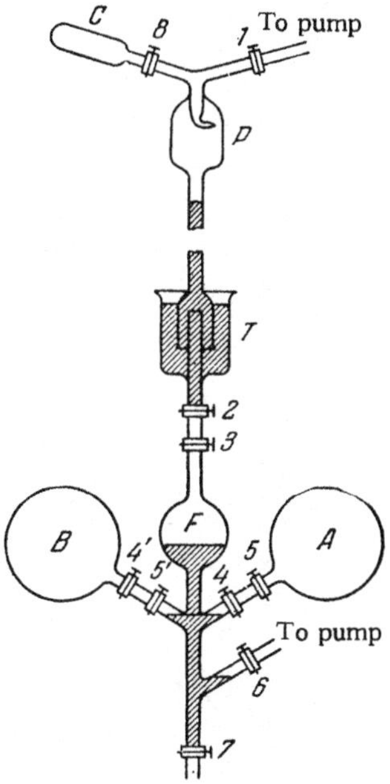

FIG. 28. Hansen's apparatus for the preparation of standard mixtures.

are made with a manometer. Since the individual volumes of all the gases taken are equal, the ratio of their concentrations in the mixture will be equal to the ratio of the pressures used in filling *C*. When displacing a gas from chamber *F*, the latter must be completely filled with mercury so that no gas bubble will remain inside (the gas can be completely forced out of chamber *F* and into receiver *C* only if the mercury rises high enough to fill the orifice of stopcock *3*; however, this may result in contamination of the mercury with stopcock grease).

In Hansen's apparatus [295] the standard mixture receiver *C* is placed above a barometric leg (Fig. 28). The preparation of a standard mixture proceeds as follows. All stopcocks are closed and chamber *T* is filled with mercury. Then stopcocks *1*, *8*, *6* and *3* are opened and vessels *C* and *F* are evacuated. The mercury now rises in the barometric leg. Stopcocks *1* and *6* are then closed, *4* and *5* are opened, and gas from cylinder *A* is allowed to flow into vessel *F* and the space between stopcocks *2* and *3*. Stopcocks *3*, *4* and *5* are then closed, and stopcock *2* is opened. The mercury then fills the tube *2-3*, forcing the gas into vessel *C* through trap *P*. Stopcock *2* is then closed while *3* is opened; the mercury drains into vessel *F*, and fills the space on top of stopcock *7*. The space *2-3* is refilled with gas from *F*, and the operation is repeated. In this manner a portion of the gas, or all of it, can be transferred to vessel *C*.

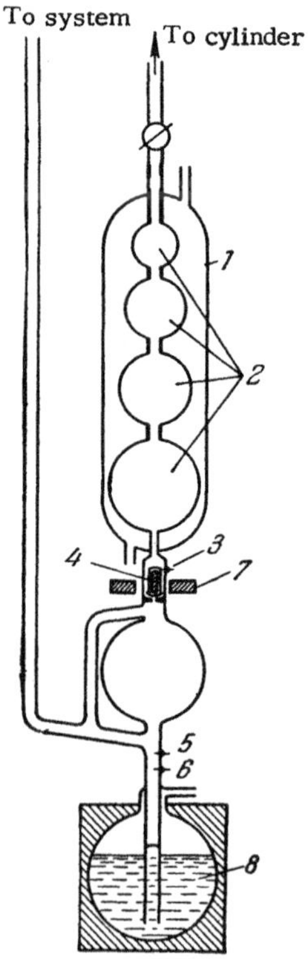

FIG. 29. Töpler pump with automatic mercury control. 1—Water cooling; 2—measuring vessels; 3, 5, 6—tungsten input leads; 7—solenoid; 8—mercury.

Before admitting the next mixture component, the mercury is drained from vessel *F* and the connecting tubes via stopcock *7*. The entire system, (except container *C*) is evacuated via stopcocks *1* and *6* (while stopcocks *2*, *8*, *4* and *7* remain closed). The subsequent steps are repeated in the same order as before. The Hansen installation does not, however, provide for automatic raising and lowering of the mercury level.

Roberts and Madison [296] described a Töpler pump system with automatic control of the mercury level (see Fig. 29). Two solenoids, one in the vacuum section and the other in the section communicating with the atmosphere, are operated by means of three tungsten leads welded into the glass. The gas is transferred into the space located above the pump via valve *4*. The transfer is achieved by means of a relay circuit, each cycle lasting half a minute. This system is capable of nearly complete gas transfer. The automatic control and elimination of greasing render this arrangement particularly desirable for many uses.

The advantage of preparing standard mixtures in gas transfer systems is associated with the fact that the initial pressure in cylinders *A* and *B* may be lower than the pressure required for container *C*. Furthermore, such systems allow preparation, without much difficulty, of standard mixtures consisting of any desired number of components.

It is necessary to bear in mind that the mixing of gases proceeds rather slowly. The prepared standard should therefore be left standing before it is used. The rate of mixing can be determined experimentally.

We shall next describe equipment for preparation of very dilute gas mixtures [297]. This arrangement can be used effectively when the minor component can be readily liquefied. The apparatus (see Fig. 30) consists of two interconnected vessels. A stream of gas flows through the upper vessel; the lower one is filled with the liquefied gas and the space above it is saturated with the vapor (gas). For a given liquid, the rate of diffusion depends on the pressure of the flowing gas as well as the pressure of the saturated vapor (i.e., on the temperature of the liquid). The diffusion rate can be calculated from theory; thus, the composition of the mixture can be computed if the gas flow rate is known. The method can be used, for instance, in the preparation of standards for determining mercury vapor concentrations in various gases.

In analytical research work it is best to prepare the standards directly in the discharge tube; the techniques which have been described above also apply in this case. However, a binary mixture may be prepared in the discharge tube in a simpler fashion: an accurately known quantity (usually the volume contained within the space between two stopcocks) of the principal component of the mixture is admitted into the tube, and the pressure in the discharge tube is measured. Then one calculates the pressure at which the minor component must be present in the tube (when alone) to give the desired partial pressure in the mixture. The tube is now evacuated and the minor component is introduced. Its pressure is checked, the primary component is admitted, and the pressure is rechecked (the pressure is thus determined twice—after the first and then after the second components of the mixture have been admitted into the tube). Before starting the analysis, some time must be allowed for the mixing of gases. At pressures of the order of a few mm Hg, the required waiting period does not exceed 5

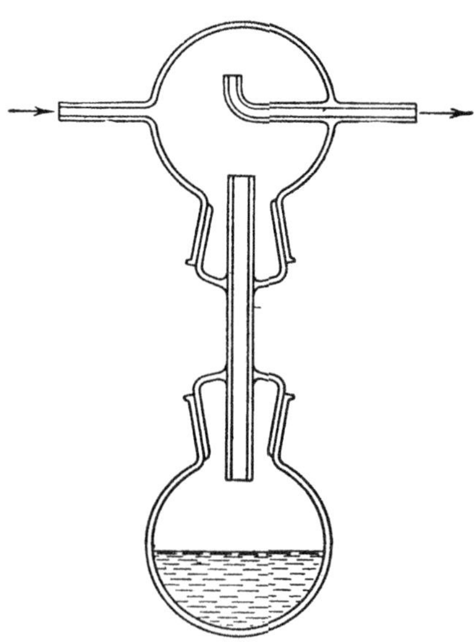

FIG. 30. Apparatus for the preparation of mixtures containing a readily liquefied gas as an admixture.

minutes. The required time for mixing can be readily determined by checking the time from the start of the mixing to the point at which the ratio of intensities of the lines of the components in the mixture becomes stable. In large-diameter discharge tubes, mixing is completed within a few seconds.

Mixture preparation is also simplified by metering (dosimetric) stopcocks. If electrically operated stopcocks (see Section 7) are available in the system, then the gas quantities can be measured by means of the time the valve is open (provided pressure is constant).

All the methods described above are suitable for the preparation of standards in which the minor component is present in concentrations of several hundredths of one percent, or higher. Standards with a lower content of the minor component should be prepared by the dilution method.

Various procedures have been developed for preparing standard mixtures by the flow method. One such procedure was proposed by Hersch and Whittle [298] and the equipment is shown in Fig. 31.

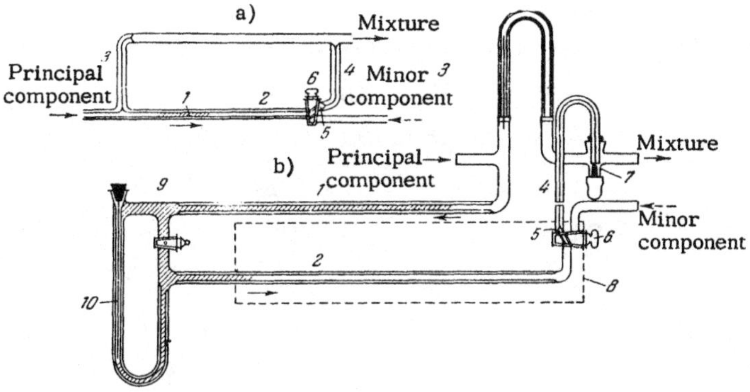

FIG. 31. Apparatus for preparing standard mixtures by the flow method.

The key part of the device consists of a syringe filled with a viscous liquid acting as a piston *1*. Because of the presence of constriction *3*, the liquid piston moves at a rate proportional to the flow rate of the principal component of the mixture and therefore the concentration of the minor component in the resulting mixture is not affected by flow rate fluctuations.

Figure 31a shows the basic variant of the method. The minor component is alternately introduced into the burette *2* through *5*, or is discharged into the stream of the principal gas component through capillary *4*. The direction of this flow depends on the position of stopcock *6*. The dead space should be minimum since it causes errors in preparing the mixture.

In a more complex version of the same idea (Fig. 31b) the desired composition of the mixture is reached instantaneously. A three-way stopcock *6* is connected to a mercury cutoff *7*. Burette *2* in this case has a larger capacity and its filling is accelerated since the viscous liquid acting as a piston can be returned to the burette via the auxiliary route *9* (there are no capillaries in this channel). The contamination of the gas depends on the nature of the piston liquid, the length of the nickel tube inserted into the U-shaped tube *10*, and the length of the wire running through capillary tube *3*. For precision work, the temperature inside the apparatus must be kept constant (thermal lagging *8*).

This device was used by Hersch and Whittle for preparing a mixture of nitrogen and oxygen ($6 \cdot 10^{-6}$% oxygen) at a flow rate of 100 ml/min. Even lower concentrations of the minor component can be obtained by mixing two streams—one of pure gas, the other containing the impurity at a known concentration.

A device in which several gas streams can be mixed in any desired proportion was described by Ageykin et al. [299].

Where standard mixtures must be stored for a long time one must consider the effects of gas adsorption on, or gas evolution from, the glass surface of the storage vessel (the ratio of components of the mixture must stay constant if the mixture is to serve as a standard). These effects become particularly important if the standards and samples are stored at low pressures. A number of studies [300-305] have shown that gas adsorption on glass and quartz depends on the condition of the surface, the temperature, and the gas pressure inside the vessel. At a constant temperature, the gas adsorption increases with pressure. The increase is fairly rapid at very low pressures, slows down in the higher pressure range, and finally becomes independent of pressure. At constant pressure, adsorption decreases with increase in temperature. Spectroscopic studies on nitrogen—oxygen mixtures stored in degassed glass vessels showed that after one month such mixtures became progressively depleted in oxygen, the loss being most pronounced at pressures of 1 mm Hg and lower.

One method for searching for leaks in gas discharge system involves discharging a spark from a Tesla coil. It has been noted that this produces a copious evolution of CO_2 and water vapor from the tube walls, even if the latter have been thoroughly degassed. Apparently, such high frequency currents knock out gas molecules that have dissolved in the glass.

Thus, it is recommended that standard mixtures be prepared and stored at pressures above 100 mm Hg. Low pressure gas samples should be analyzed as soon as possible because the original composition of the sample may change on storage. The Tesla coil should not be discharged onto the glass surfaces of the gas sample container, the walls of the connecting tubing in the discharge system, or the discharge tube itself.

10. PREPARATION AND PURIFICATION OF GASES

Most of the relatively small quantities of gases needed for compounding standard mixtures can be prepared in the laboratory

[276-278]. Thus, hydrogen and oxygen are obtained by electrolysis of a 10% alkali solution. To remove water vapor, hydrogen and oxygen may be passed through a long tube filled with phosphorus pentoxide, and then through a coil placed in a Dewar flask containing liquid nitrogen. Traces of oxygen may be removed from hydrogen by passing the latter through a quartz tube filled with clean copper shavings and kept in a furnace at 650-700°C. Traces of hydrogen are removed from oxygen by passing the latter through a quartz tube filled with cupric oxide and kept in a furnace at 550-600°C. To decompose traces of ozone, the oxygen is passed through a tube containing active manganese dioxide (hopkalite).

Spectroscopically pure hydrogen can be obtained from a burner gas jet by permitting the hydrogen to diffuse through a palladium capillary welded to a vacuum-producing system and heated to incandescence by a burner. Spectroscopically pure oxygen is obtained by decomposition of potassium permanganate. The traces of CO_2 are removed by passing the oxygen through a trap filled with solid potassium hydroxide, or by freezing out with liquid nitrogen. Small quantities of oxygen can also be obtained from the air, by means of diffusion through a silver tube. In this procedure, the rate of oxygen accumulation in a 1 liter vessel was 2 mm^3/hr using a 3 mm I.D., 150 mm long tube with a wall thickness of 0.25 mm heated to 700°C [306]. Pure nitrogen is obtained by decomposition of sodium, potassium or barium azide (heating in vacuo). Nitrogen can also be obtained from sodium nitrite, by a "wet" method, according to reaction:

$$NaNO_2 + NH_4Cl \rightarrow NaCl + N_2 + 2H_2O.$$

A vacuum flask fitted with a dropping funnel is one-third filled with a saturated solution of ammonium chloride, the flask is heated, and a concentrated sodium nitrite solution is added dropwise through the funnel. The reaction is very vigorous. The nitrogen thus obtained contains traces of oxides, which are removed by passing the gas over red-hot copper shavings.

Carbon monoxide is obtained from formic acid which is decomposed by sulfuric acid. Carbon dioxide is obtained by the decomposition of neutral magnesium carbonate ($MgCO_3$) in vacuum at 400-700°C.*

The inert gases are prepared by fractional distillation of air or natural gas. They are difficult to prepare using limited laboratory facilities, and therefore such gases are bought from large manufacturers. Pure inert gases are marketed either in metal cylinders or in sealed glass containers. To draw small gas samples from a glass container, one attaches to the latter two interconnected stopcocks, or the valve shown in Fig. 18. A small glass- or Teflon-coated iron bar is placed inside a glass tube which connects the stopcock and the sealed end of the storage container. Before the sealed end is broken, the connector tubing is thoroughly evacuated and degassed, the stopcock is closed, and the sealed tip is broken off by moving the iron bar by means of a magnet.

The inert gases may be purified in a discharge tube filled with potassium or sodium vapor, or through absorption of the impurities by lithium or calcium at high temperatures. Helium is freed of all impurities (including inert gases) in a trap containing activated carbon at the temperature of liquid nitrogen (-196°C).

Bennet [307] recommends the use of glass vacuum pumps for obtaining high-purity inert gases. A glass pump offers many advantages since it will cause no contamination, is easily heated to remove adsorbed gases and performs well over a wide

*The exact decomposition temperature of $MgCO_3$ will vary depending on impurities.

range of pressures. It has few moving parts and its suction rate is high (2 liters/min at p = 1 atm).

Traps filled with natrolite, kieselguhr, diatomite or kaolin, all of which are natural adsorbents for hydrocarbons, can be used to free an inert gas from hydrocarbon impurities. Somewhat less effective adsorbents for hydrocarbons are activated carbon and silica gel (the adsorption of hydrocarbons and their mixtures on carbon and silica gel is described in [308]).

The removal of CO_2 and water vapor from air is achieved with the aid of alumina gel. Traps filled with alumina gel can be repeatedly regenerated by calcination.

11. LIGHT SOURCE POWER SUPPLIES FOR SPECTROSCOPY OF GAS MIXTURES

In spectral gas analysis the power needed to produce the excitation is supplied by DC and AC generators as well as h-f (r-f) oscillators.

Thus, a DC glow discharge will occur in a discharge tube when the latter is supplied by an 0.5-1 kW rectifier. The wiring diagram of the full-wave rectifier is shown in Fig. 32. The circuit includes a filter consisting of a capacitor (C = 10-20 μF) and a choke (inductance $L \approx 20$ henries). The rectified current is regulated by means of two variable resistors, one of which is connected

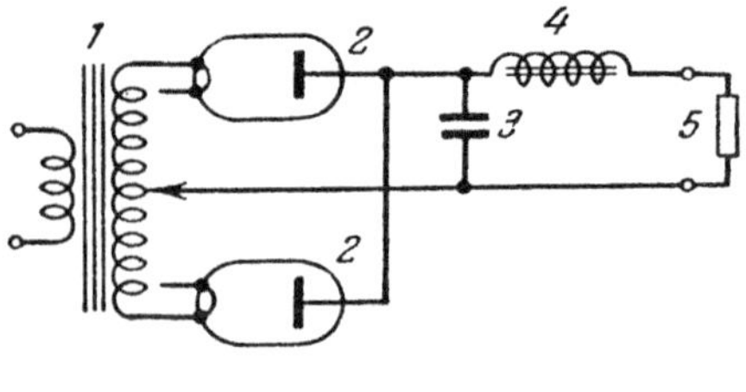

FIG. 32. Full-wave rectifier. 1—Transformer; 2—vacuum tube rectifiers; 3—capacitor; 4—choke; 5—load.

to the primary of the transformer, and the other hooked up in series with the discharge tube. The resistance of the last resistor should be of the same order as that of the discharge tube. Both the resistor and milliammeter must always be inserted in the grounded part of the circuit. Either the positive or the negative terminal of a rectifier tube can be grounded. The rectifier output voltage generally amounts to 1000–1500 V.

Such a high-voltage rectifier is also suitable for feeding a pulse generator (see Fig. 13). In this case, the rectifier voltage should be increased to 3000 V for best results. A high capacitor charge, which sets up a potential higher than the breakdown voltage of the discharge tube, is obtained by connecting a spark gap or a thyratron in series with the tube. The condenser capacitance can be varied from 2–3 to 100 μF, depending on the required flash intensity. Resistor R inserted in the capacitor charging circuit makes it possible to achieve a pulsed mode of operation while working with relatively low capacitance values. This resistance increases the charging time of the capacitor so that flashes (pulses) can be separated. In cases where the firing potential of the discharge is very low, a thyratron or spark gap must be used, because otherwise the capacitor is unable to discharge unless its potential is quite high.

If the discharge tube is powered directly from the line, then the voltage must be raised to 1500–3000 V by means of a step-up transformer.

In spectroscopy of gas mixtures high-frequency discharges are often used to provide a source of light. Three types of h-f oscillators, which are typical of those used in this application, have been developed by the experimental workshops at the Physical Research Institute of Leningrad State University. These are the VG-2 and VG-3 types and a low power device using a double pentode. A schematic diagram of the 6-Mc, 350-W VG-2 generator is given in Fig. 33. The design provides no facilities for variation of the frequency.

In this generator, the oscillator circuit is coupled to the discharge tube through a specially designed coil whose ends are connected to the external electrodes of the tube. The discharge tube current can be varied either by changes in the coupling coil or by regulating the current flowing through the primary of the high-voltage rectifier. To achieve a steady oscillator performance the voltage supply to it must be stabilized. Fluctuations in the AC line current may lead to serious errors in the analysis.

Generators of the VG-2 type can be used only when high analytical accuracy is not required and where errors of 10-15% are acceptable. A more stable oscillator that will ensure reproducibility of results within 5-7% (even with an unstabilized line voltage) is the low power, push-pull oscillator of the type shown in Fig. 34 [310]. This particular generator uses a capacitance-coupled double pentode. This arrangement achieves a stable peak voltage, whereby the effect of the discharge tube on the oscillator frequency and the output voltage is reduced. The frequency of this particular 35-40 W oscillator can be varied within 58-60 Mc by means of a variable capacitor C_1. The oscillator circuit is based on capacitance C_1 and inductance L_1 and is connected through coupling capacitors C_2 and C_3 to the discharge tube. The feedback voltage is tapped from choke L_1 and fed via capacitors C_4 and C_5 to the control grids of the double pentode. The optimum coupling value is determined experimentally by varying the number of turns in the coupling coil L_1. The AC component in the control grid circuit is isolated by means of h-f chokes L_3 and L_4. The control grid bias is created by the grid current passing through resistor R_1. The screen grids are fed from the B supply through voltage-dropping resistor R_2.

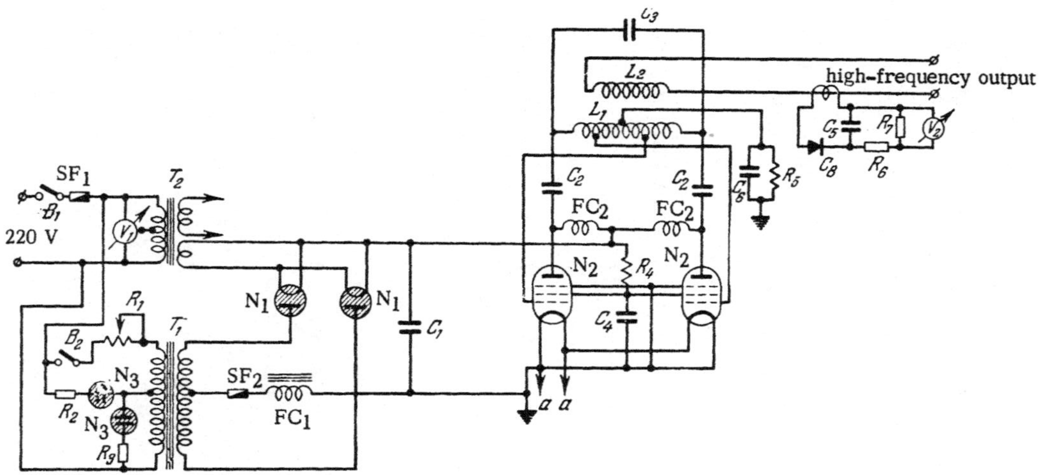

FIG. 33. The experimental high-frequency generator VG-2. C_1—1500 V, 4μF paper capacitor; C_2—2500 V, 500 pF mica capacitor; C_3—70 pF air capacitor; C_4—1000 V, 3000 pF mica capacitor; C_5—2500 V, 1000 pF sealed paper capacitor; C_6—1000 V, 5000 pF mica capacitor; V_1—AC voltmeter, 250 V; V_2—hot wire ammeter; R_1—variable resistor, 6-8 A, 50-100 kΩ; R_2, R_3—0.25 W, 15 kΩ composition resistors; R_4—1000 W, 10 kΩ vitrified resistor; R_5—same, 50 W; R_6—0.5 W, 1 kΩ composition resistor; R_7—same, 2 kΩ; N_1—hot cathode, gas-filled tube rectifiers, Soviet type VG-129; N_2—pentode, Soviet type G-471; N_3—neon pilot light; L_1—oscillator circuit coil; L_2—coupling coil; B_1—filament circuit switch; B_2—plate circuit switch; SF_1—10 A safety fuse; SF_2—same, 1 A; FC_1—filter choke; FC_2—high-frequency choke; T_1—power transformer; T_2—filament circuit transformer; C_8—copper-oxide rectifier.

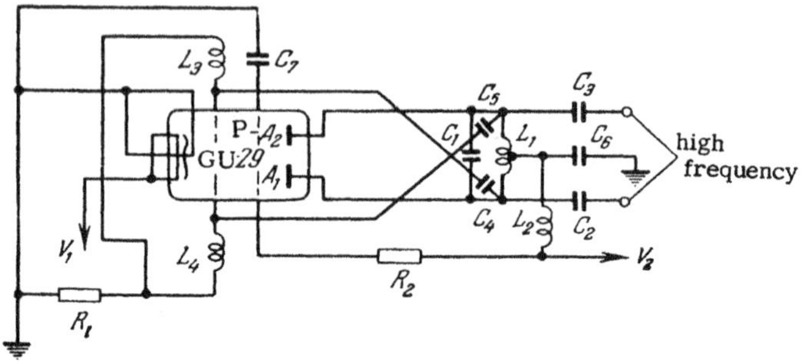

FIG. 34. The experimental low power high-frequency generator. C_1—200 pF; C_2, C_3—25 pF; C_4, C_5—50 pF; C_6, C_7—600 pF; R_1—18 kΩ; R_2—90 kΩ; V_1—6.3 V; V_2—400-700 V; L_1—0.4-2 microhenries; L_2, L_3, L_4—chokes; P—double pentode, Soviet type GU-29.

This type of generator may be used as an excitation source in conventional gas analyzers (see Section 14).

When it is desired to analyze mixtures of gaseous isotopes, oscillators such as the VG-3 (Fig. 35) are used [311]. In addition to stable performance, this oscillator is capable of high power output, both of which are necessary. Improved stability of operation is achieved by means of an electron-optical feedback. Thus, a fraction of the luminous flux emerging from the discharge tube is made to impinge on a single-stage photomultiplier. The electric signal generated in multiplier is amplified by an amplifier in the feedback circuit of the oscillator and is fed, in the form of a modulating signal, to the h-f oscillator circuit. Thus the power output of the oscillator varies with the luminous flux in the load circuit. This method of stabilizing the luminous flux permits compensation of the variation in brightness of the glow caused by the instability of the line voltage. Since the oscillator is multi-stage, the gain required in the feedback circuit may be substantially reduced.

Power control is achieved in the low power stage which uses tube N_2 operating without grid currents. As a result, a low gain DC amplifier can be used in the feedback circuit.

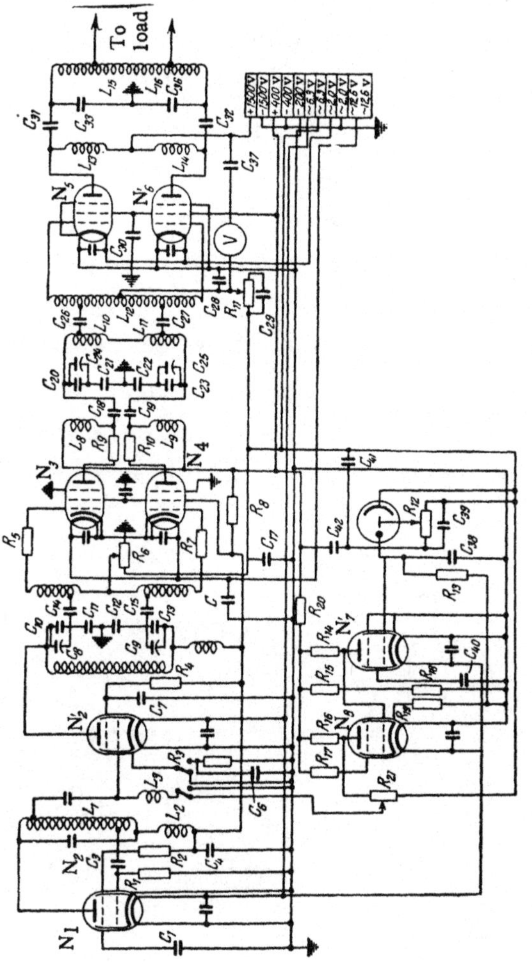

FIG. 35. The experimental high frequency oscillator, type VG-3.

The master oscillator (N_1) is hooked up in a conventional Hartley oscillator (coil-coupled) circuit using the same tetrode as in the power control stage.* The amplification stages of the high-frequency oscillator operate in a push-pull mode and use Soviet tubes GU-29 (N_3, N_4) and GK-71 (N_5, N_6).

The feedback amplifier uses directly coupled tubes N_7 (Soviet type 6Zh8) and N_8 (Soviet type 6P6). Feedback level control is achieved by means of a potentiometer.

A stabilized rectifier supplies power to the photomultiplier, the master stage, the bias circuits of the h-f oscillator, as well as the DC amplifier. This particular generator is tuned to a frequency of 1.6 Mc and delivers an output of 300 W.

Microwave frequency oscillators, which offer many advantages over low frequency generators (see Section 5), are now finding increasing use in spectral analysis. A schematic diagram representing one of the many possible designs of a magnetron oscillator [241]

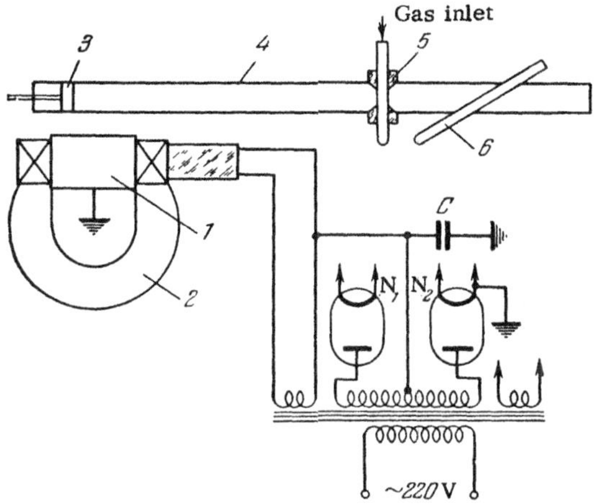

FIG. 36. Magnetron oscillator. 1—Magnetron; 2—magnet, 1200 gauss; 3—noncontacting plunger; 4—waveguide; 5—discharge tube; capacity C = 4 μF; 6—nonreactive load.

*Soviet type 6P6.

is shown in Fig. 36. The magnetron, designed for continuous operation at 3000 Mc, has the following parameters: U_{plate} = 4150 V; i_{plate} = 50-70 mA; $U_{filament}$ = 6.3 V; H = 1200 gauss. The discharge tube is installed inside a waveguide, in the antinode of the electromagnetic wave. The light emanating from the tube is extracted through a slit in the side wall of the waveguide. Adequate protection against microwave-frequency radiation should be provided for the personnel working with these oscillators.

Several papers on spectroscopy and spectral analysis [240, 243, 312] mention the use of highly stable standard oscillators operating at 2450 Mc and giving a power output of 125 W.

12. THE SPECTROSCOPIC SYSTEM

The primary function of the spectroscopic system, as far as gas analysis is concerned, is to disperse the radiation emitted by the glowing gaseous mixture and to resolve it into its monochromatic components. The operating principle of the spectral instruments used for this purpose is based on light dispersion, diffraction or interference. Spectra may be observed visually, they may be photographed, or they may be recorded by means of a photoelectric radiation receiver-recorder. An assembly combining a monochromator with a photoelectric attachment consisting of a recording radiation detector is called a spectrometer.

The basic components of any spectroscopic device—the slit, the dispersing system and the optical system proper—all serve the purpose of obtaining spectral lines, of which each is a monochromatic image of the slit.

The dispersing system may be a prism, a diffraction grating or a set of interferometer plates. The many existing types of prism instruments use glass, quartz or fluorite prisms of various shapes, depending on the spectral region under consideration. The use of prism instruments is limited to the visible and the relatively near UV regions.

The introduction of ruled diffraction gratings has greatly improved spectroscopic equipment. Instruments incorporating diffraction gratings are finding an ever increasing use, especially in vacuum ultraviolet spectroscopy, as well as in some types of analysis of isotopic mixtures.

We shall omit here the detailed description of the various available spectroscopic instruments. A review of these can be found in any of the general manuals on spectroscopy and spectral analysis [25, 208, 209, 313-317],* as well as in the literature available from the manufacturers. We shall, however, discuss interference filters in some detail (see Section 13), principally because the handbooks generally devote little space to these parts of the apparatus.

The basic parameters describing any spectroscopic instrument are its dispersion, resolving power and light transmittance.

Dispersion

A distinction should be made between linear and angular dispersion. The *angular dispersion* of a spectrometer defines the angle which separates two beams whose wavelengths differ by $\Delta\lambda$ after they pass through the beam splitter. The angular dispersion is given by

$$D_\varphi = \frac{d\varphi}{d\lambda}, \tag{3.4}$$

where $d\varphi$ is the angular separation of the beams in the dispersing element of the system.

At minimum deviation and for a single-prism instrument

$$D_\varphi = \frac{d\varphi}{d\lambda} = \frac{2\sin\frac{\alpha}{2}}{\sqrt{1 - n^2\sin^2\frac{\alpha}{2}}}\frac{dn}{d\lambda}, \tag{3.5}$$

*Jacquinot [318] gives a very interesting and original description of the various aspects of spectroscopic techniques.

where α is the refraction angle of the prism, n is the refractive index and the factor $dn/d\lambda$ is specific to the prism material.

In instruments using a diffraction grating, the angular dispersion is given by

$$D'_{\varphi} = \frac{d\varphi'}{d\lambda} = \frac{k}{d \cos \varphi'} = \frac{\sin \varphi' + \sin \psi}{\lambda \cos \varphi'} \tag{3.6}$$

where d is the grating constant (the distance between two adjacent ruled lines), φ' is the angle of diffraction and ψ is the angle of incidence of the light impinging on the grating.

Two light rays of wavelengths λ_1 and λ_2, resolved by the splitter and made to diverge at an angle $d\varphi$, will project the image of two lines onto the focal surface of the instrument. If the distance between the lines on the focal plate is dl, then

$$\frac{dl}{d\lambda} = D_{\lambda} \tag{3.7}$$

where the factor D_{λ} is the *linear dispersion* of the instrument; it is usually given in terms of its reciprocal (or Å/mm). The linear and angular dispersions are related by

$$D_{\lambda} = D_{\varphi} F = \frac{d\varphi}{d\lambda} F, \tag{3.8}$$

where F is the focal length of the focusing (objective) lens.

Linear dispersion can be increased either by increasing the angular dispersion or the focal length of the focusing system. To improve the angular dispersion of a prism instrument one must either increase the angle of refraction of the prism or the number of prisms. Alternatively, one can use a prism made of some material with better dispersive properties.

The $dn/d\lambda$ value is a decreasing function of the wavelength. It follows that prism instruments have lower dispersions in the infrared than the ultraviolet portion of the spectrum.

Within the limits of one order, the linear dispersion of a grating instrument may be assumed virtually constant, provided the angle of incidence of the light does not deviate too far from the normal. It is seen from Eq. (3.6) that if the angle φ' is small, then $\cos \varphi' \sim 1$ and dispersion is constant. If the ray falls obliquely upon the grating, the value of φ' may reach several dozens of degrees so that dispersion will vary with the wavelength (as in the case of prisms, dispersion is greater in the short than it is in the long wavelength region).

Resolution

Another essential characteristic of any instrument is its *resolving power*, which is given by

$$R = \frac{\lambda}{\delta\lambda}, \tag{3.9}$$

where $\delta\lambda$ is the smallest wavelength difference between two equal intensity lines which can still be resolved by the instrument operating at a given wavelength λ.

A spectral line is actually the slit image which is broadened by diffraction. If the slit is very narrow, the width of the image will depend almost solely on the width of the zeroth order diffraction maximum. Two lines are considered (see Fig. 37) resolved if the diffraction minimum of one aligns with the diffraction maximum of the other (*Rayleigh's criterion*). In that case the drop in intensity between the two maxima is 20%.

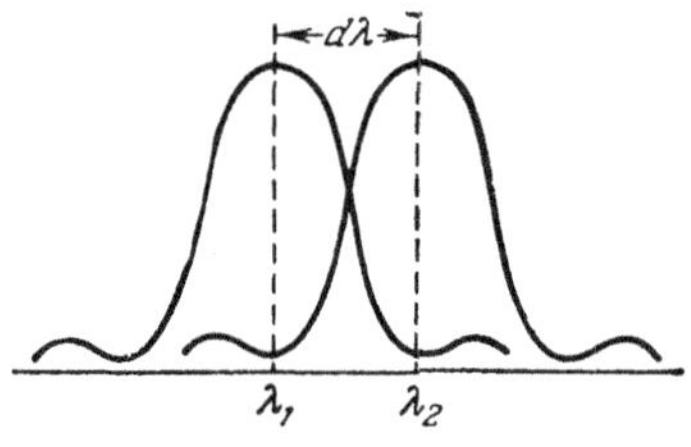

FIG. 37. Criterion for evaluating the resolution of two spectral lines.

Two lines differing markedly in intensity may not be resolved by the instrument even though their relative position satisfies the Rayleigh criterion. On the other hand, when the accuracy in the measurement of intensity is considerably higher than 20% (as it is in photoelectric recording), the instrument may be able to resolve lines separated by a distance which is smaller than that required by Rayleigh's criterion.

The resolving power of a prism instrument is determined by resolution of which the prism is capable when completely illuminated:

$$R = a\frac{dn}{d\lambda}, \qquad (3.10)$$

where a is the thickness of the base of the prism and $dn/d\lambda$ is the characteristic dispersion specific to the prism material.

In the same way the resolving power of an instrument using a diffraction grating depends on the resolving power of the grating.

The theoretical resolving power of a grating is given by

$$R = kN, \qquad (3.11)$$

where N is the number of illuminated ruled lines in the grating and k is the order of spectrum.

The actual resolving power of a spectroscopic instrument is considerably lower than the theoretical. It is affected by such factors as imperfections in the optical system, light diffraction by the diaphragms, stray light, and the width of the slit. It turns out that below the so-called "normal" width of the slit, the width of the line image on the focal plane remains constant regardless of whether the width of the slit is further reduced. This means that reduction of the width of the slit to below normal does not improve the resolving power of the instrument. The *normal width of the slit* depends on the wavelength:

$$S = \frac{\lambda f}{d}, \qquad (3.12)$$

where f is the focal length of the collimator lens and d is the lens diameter.

Sometimes it is not possible to fully utilize the available resolving power of the instrument. This happens with spectrographs, where the limiting factor is the size of the grain of the photographic plate.

Light transmittance

The efficiency with which a spectroscopic device utilizes the incident radiation is governed by the *light transmittance* of the system. If the condenser lens is large enough to transmit all of the incident light, then the luminous flux dF entering the spectrometer will be a function of the brightness b_λ of the light source, the area ΔS_1 of the entrance slit, and the solid angle $d\Omega$:

$$dF = b_\lambda \, d\lambda \, \Delta S_1 \, d\Omega \tag{3.13}$$

or, considering that

$$d\Omega = \frac{\pi}{4}\left(\frac{D_1}{F_1}\right)^2,$$

$$dF = \frac{\pi}{4}\left(\frac{D_1}{F_1}\right)^2 b_\lambda \, d\lambda \, \Delta S_1, \tag{3.14}$$

where D_1 is the diameter of the collimating objective and F_1 is its focal length.

The light transmittance of an instrument varies with the type of recording employed. The commonly used photographic recording methods measure illuminance produced by the light source in the focal plane of the instruments, integrating it over time to yield ***exposure***, whereas the photoelectric detector measures only the instantaneous light flux transmitted by the optical system to the photosensitive layer of the radiation receiver. Let us see which parameters of the instrument are controlling in each case.

1. The case of a line spectrum

a) *Photoelectric detection (in terms of light flux).* A fraction of the flux entering the instrument is absorbed in the optical system, another fraction is lost by reflection, and still another fraction is scattered. The luminous flux emerging from the instrument can thus be expressed as dF':

$$dF' = \alpha\, dF = \frac{\alpha\pi}{4}\left(\frac{D_1}{F_1}\right)^2 b_\lambda\, d\lambda\, \Delta S_1, \qquad \alpha < 1. \tag{3.15}$$

If the exit slit is able to pass the complete image of a line, the emerging total flux is

$$F' = \frac{\alpha\pi}{4}\left(\frac{D_1}{F_1}\right)^2 \Delta S_1 \int_0^\infty b_\lambda\, d\lambda, \tag{3.16}$$

But $\int_0^\infty b_\lambda\, d\lambda$ is equal to the total intensity (brightness) of the line B.

Thus

$$F' = \frac{\alpha\pi}{4}\left(\frac{D_1}{F_1}\right)^2 B\,\Delta S_1. \tag{3.17}$$

It follows that the light transmittance (F'/B) of the spectral instrument, given by

$$\frac{F'}{B} = \frac{\alpha\pi}{4}\left(\frac{D_1}{F_1}\right)^2 \Delta S_1, \tag{3.18}$$

is proportional to the transmittance (relative aperture number) of the collimator and the area of the entrance slit ΔS_1.

b) *Photographic recording (in terms of illuminance).* Let $\Delta S_1'$ designate the area of the image of the entrance slit produced by the optical system of the instrument. If $\Delta S_1'$ is considerably greater than the natural width of the line then the illuminance E in the focal plane of the instrument is

$$E = \frac{\Delta F'}{\Delta S_1'} = \frac{\alpha\pi}{4}\left(\frac{D_1}{F_1}\right)^2 B\,\frac{\Delta S_1}{\Delta S_1'}. \tag{3.19}$$

If the camera and collimator lenses are of equal diameter ($D_1 = D_2$), produce identical magnification $q = 1$, and have the respective focal lengths F_1 and F_2, then

$$\frac{\Delta S_1}{\Delta S_1'} = \left(\frac{F_1}{F_2}\right)^2$$

in which case the illuminance expression (3.19) can be rewritten as

$$E = \frac{\alpha\pi}{4}\left(\frac{D_2}{F_2}\right)^2 B. \tag{3.20}$$

The light transmittance of the spectral instrument thus becomes

$$\frac{E}{B} = \frac{\alpha\pi}{4}\left(\frac{D_2}{F_2}\right)^2 \tag{3.21}$$

that is, it is proportional to the transmittance (relative aperture number) of the objective of the camera and is independent of the entrance slit area.

2. *The case of a continuous spectrum*

a) *Photographic recording (in terms of illuminance).* The illuminance produced by a portion of a continuous spectrum in the focal plane of the instrument is

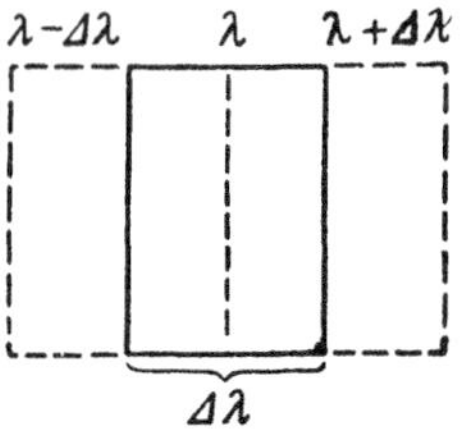

FIG. 38. The area of overlap of slit images.

$$E = \frac{\alpha\pi}{4}\left(\frac{D_2}{F_2}\right)^2 \int\limits_{\Delta\lambda} b_\lambda \, d\lambda. \tag{3.22}$$

Here $\Delta\lambda$ is the arc of overlap of images S_1 of the slit at several wavelengths (see Fig. 38); it is given by

$$\Delta\lambda = \frac{1}{D_\varphi F_2} S_1', \tag{3.23}$$

where D_φ is angular dispersion of the system and S_1' is the width of the image of the entrance slit. If the camera and collimator lenses are of equal diameter ($D_1 = D_2$) and produce an identical magnification $q = 1$, then

$$S_1' = S_1 \frac{F_2}{F_1}, \tag{3.24}$$

where S_1 is the width of the entrance slit and

$$\Delta\lambda = \frac{1}{D_\varphi F_1} S_1. \tag{3.25}$$

Then

$$\int_{\Delta\lambda} b_\lambda \, d\lambda = \bar{b}_\lambda \, \Delta\lambda = \bar{b}_\lambda \frac{S_1}{D_\varphi F_1}, \tag{3.26}$$

where $\bar{b}_\lambda$ is the mean value of the brightness distribution function in the range $\Delta\lambda$.

The illuminance expression (3.20) then becomes

$$E = \frac{\alpha\pi}{4} \left(\frac{D_2}{F_2}\right)^2 \frac{\bar{b}_\lambda S_1}{D_\varphi F_1}. \tag{3.27}$$

Thus, the transmittance of the spectral instrument $E/\bar{b}_\lambda$ which has the form

$$\frac{E}{\bar{b}_\lambda} = \frac{\alpha\pi}{4} \left(\frac{D_2}{F_2}\right)^2 \frac{S_1}{D_\varphi F_1} \tag{3.28}$$

is proportional to the transmittance (relative aperture number) of the objective of the camera and the width of the entrance slit, and varies inversely with the angular disperison and the focal length of the collimator lens.

b) *Photoelectric detection (in terms of luminous flux).* The luminous flux $\Delta F'$ emerging through the exit slit of area ΔS_2 is given by

$$\Delta F' = E \, \Delta S_2, \tag{3.29}$$

or

$$\Delta F' = \frac{\alpha\pi}{4} \left(\frac{D_2}{F_2}\right)^2 \frac{\bar{b}_\lambda S_1}{D_\varphi F_1} \Delta S_2 = \frac{\alpha\pi}{4} \left(\frac{D_1}{F_1}\right)^2 \bar{b}_\lambda \frac{S_2}{D_\varphi F_2} \Delta S_1 \tag{3.30}$$

where S_2 is the width of the exit slit. The transmittance of the instrument

$$\frac{\Delta F'}{b_\lambda} = \frac{\alpha\pi}{4}\left(\frac{D_1}{F_1}\right)^2 \frac{S_2}{D_\varphi F_2}\Delta S_1 \tag{3.31}$$

is then proportional to the transmittance (relative aperture number) of the collimator lens, the entrance slit area and the width of the exit slit, while it varies inversely with D_φ and F_2.

Comparison of transmittances of prism and diffraction grating instruments shows that illuminance is about the same in the two types of spectral devices. In terms of the luminous flux, however, the diffraction grating appears somewhat superior. The greater angular dispersion of the grating allows the use of wider slits and smaller focal lengths. As a result a greater light flux reaches the detector at the same resolution of the instrument.

The above considerations permit a rational selection of an instrument and slit dimension for specific analytical requirements. The correct choice of the instrument is particularly important where high sensitivity of analysis is required (see Section 21).

Gas spectra show fewer lines than metal spectra. Therefore it is usually not necessary to use high-dispersion instruments (except in spectral analysis of gaseous isotope mixtures). In some cases, it is even possible to use monochromator filters (which provide a dispersion far lower than that of even the crudest spectral device) without reducing the analytical sensitivity (see Section 26).

Most spectroscopic methods for gases use analytical line pairs located in the visible spectrum. As a result, instruments with glass optics are quite sufficient.

Spectral instruments capable of delivering a high luminous flux are particularly valuable in photoelectric detection of spectra. In this application, interference filters are superior to dispersion devices. The light flux from an interference filter is a function of the surface area of the latter, and may thus be several dozen times greater than the flux dispersion instrument which is limited by the width of the slit.

Illumination of the slit of the spectral instrument

The light transmittance and resolving power of a spectral instrument cannot be fully utilized unless the collimator lens is completely filled with light from the entrance slit. In most cases this is achieved with the aid of special optical systems. Such systems are generally called condensers, or condensing systems.

The simplest condenser is a spherical converging lens. Its relative aperture (the ratio of the lens diameter to its focal length) must be such that the angle $\omega \geqslant \omega'$ (see Fig. 39) so that the collimator is completely filled with light. The required condenser magnification and its position relative to the slit should be calculated from the thin lens equation. Projecting a reduced source image upon the slit is not advisable, since this leads to a marked nonuniformity of illumination. As a rule, the light source is projected onto the slit at a 1:1 magnification. (In this case the distance between the source and the slit is $4f$, where f is the focal length of the condensing lens. A distance of $2f$ separates the lens from the slit.)

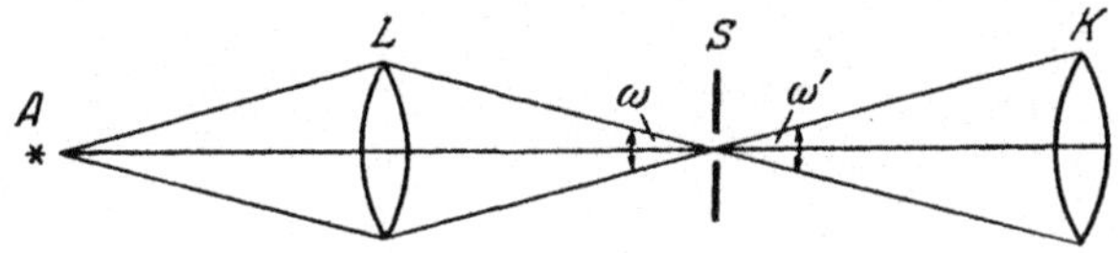

FIG. 39. Selection of the illuminating lens. A—Light source; L—illuminating lens; S—slit of the spectral instrument; K—collimator lens.

When a full-scale source image is projected upon the slit, the various points on the slit area are illuminated by rays emerging from different points in the source. A source image is then obtained in the plane of the collimator lens, with light emitted by every point in the source contributing to the slit illumination (see Fig. 40).

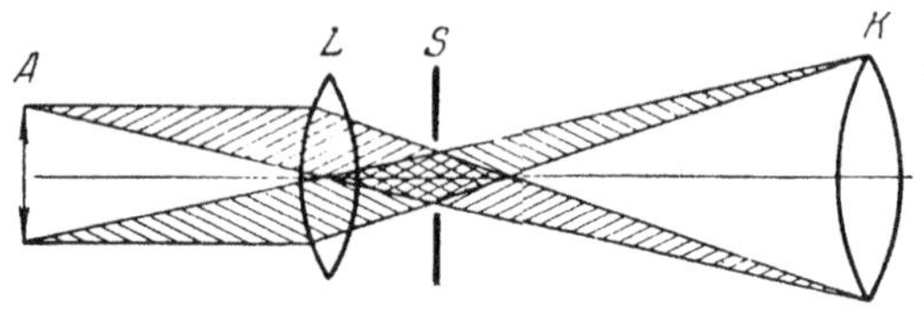

FIG. 40. Illumination of the spectrograph slit.

If the slit is illuminated by means of a single lens, the masking effect must be taken into consideration. Rays from the off-center sections of the slit are projected toward the collimator at an angle to the optical axis of the instrument, and thus may fail to reach the collimator lens. Such masking is usually eliminated by means of an adapter lens inserted in front of the slit (Fig. 41). Its focal length must be such that light transmitted by the condenser will be focused in the plane of the collimator lens. The center section of the auxiliary lens is a plane-parallel plate which does not alter the beam path, while the edges act as prisms which deflect the rays toward the optical axis.

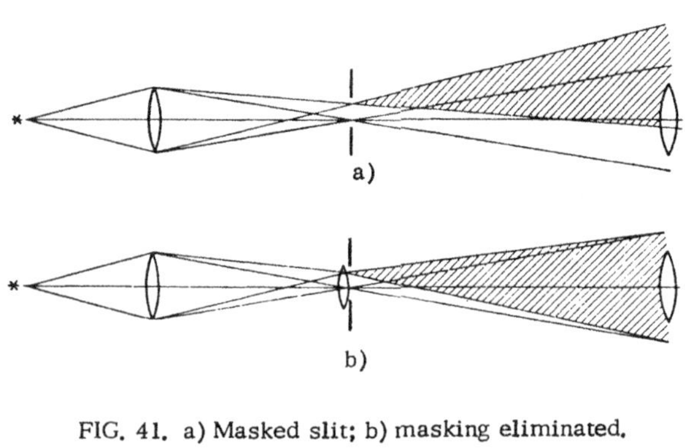

FIG. 41. a) Masked slit; b) masking eliminated.

Uniform slit illumination may be achieved with a three-lens condensing assembly, as well as special screened condensers [319].

In some cases the instrument entrance slit can be illuminated directly by the light source, eliminating the intermediate optical system. In this arrangement the source must be located on the optical axis of the instrument, at some distance from the slit. The full transmitting ability of the instrument can be utilized only if the source can be viewed through the slit at an angle wider than that at which the collimator lens is seen.

Slit illumination using various light sources, with and without condensers, is discussed at greater length in manuals on spectroscopy and spectral analysis [25, 313-317]. Still more information is available in special papers on the subject [320].

13. MONOCHROMATIC LIGHT FILTERS

It is sometimes necessary to isolate narrow spectral bands. In this case, monochromatic light filters are used. Filter passband requirements vary with the type of analysis. When it is desired to isolate widely separated lines of a line spectrum, relatively crude absorption filters (e.g., made of tinted glass) [321] or a combination of such filters are quite effective. On the other hand, isolation of a narrow band of a continuous spectrum, as well as the resolution of two closely spaced lines, requires a filter of far greater monochromaticity. This is particularly true if the line of interest is one of low intensity.

The various types of monochromatic filters now in use include

1) absorption-type light filters;
2) dispersion-type light filters [322, 323];
3) interference-polarization filters [324, 325];
4) interference-type Fabry-Perot light filters [326-328].

The reader is referred to the literature for detailed information on the first three types of monochromatic filters. In this book we shall content ourselves with the following few remarks about them.

Christiansen's dispersion light filters may be designed for the visible [329], the ultraviolet [330] or the infrared [331] range and have good optical characteristics. Thus, $T_m \sim$ 90–100%, $\delta\lambda$ = 50 Å (with up to 20 Å for twin filters). However, they are not very convenient. Their main disadvantage is that the location of the passband varies markedly with the temperature; thus, special thermostatic equipment, capable of maintaining the temperature of the filter constant within 0.1°C, is required.

The theory of these dispersion-type light filters is given in [326, 332–334].

Interference-polarization filters are very complex. Thus, for example, a filter for separation of the λ 3943 Å line of ionized calcium may consist of nine quartz lenses and 10 polarizers, the last of which is 53 mm thick [335]. The maximum transmitted light intensity does not exceed a few percent of the incident radiation. Again, the filter must be kept at a constant temperature. The principal advantage of this type of filter is its narrow transmission band (1–2 Å).

The most commonly used interference filters are of the standard Fabry-Perot type. Such filters are described, among others, by Geffcken [326], Korolev [327] and Krylova [328].

The transmission curve of any monochromatic light filter has a peak at a some wavelength (Fig. 42). The basic parameters of a monochromator are

1) the *transmission coefficient* T_m at maximum transmission (passband center) is

$$T_m = \frac{I_m}{I_0}, \tag{3.32}$$

where I_0 is the intensity of light incident on the filter and I_m is the intensity of light transmitted by the filter at the passband center $\lambda = \lambda_m$.

2) *bandwidth* $2\delta\lambda$ of the spectrum which the filter can pass;

3) *wavelength* λ_m at the center of the passband;

4) *transmission curve tails* (passband fringes) representing residual transmittance T_r in spectral regions far beyond $\lambda_m \pm 2\delta\lambda$. An alternative value often used instead of residual transmittance is the *contrast factor* of a light filter T_r/T_m;

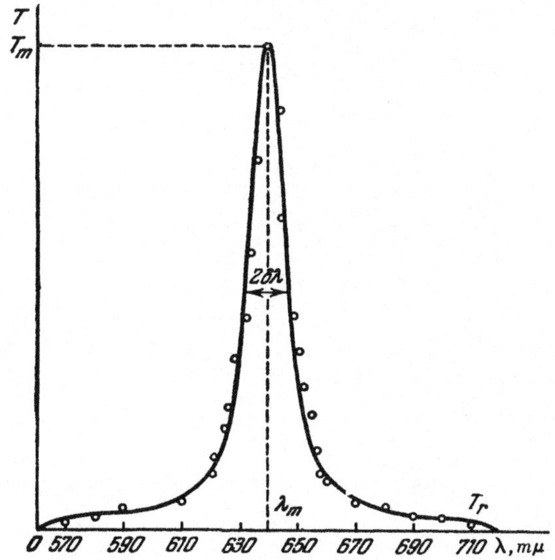

FIG. 42. Transmission curve of a monochromatic filter.

5) the *aperture* of a light filter, or the angular width of a light beam $2\delta\psi$ which the monochromator is capable of passing without a substantial increase in the spectral width of the passband. When the incident light is normal to the filter, then

$$2\delta\psi = 2n\sqrt{\frac{2}{R}}, \tag{3.33}$$

where n is the refractive index and R is the resolving power. It is obvious that the quality of a monochromatic filter improves as the values of T_m and $\delta\psi$ increase and the value of $\delta\lambda$ decreases.

Structurally, the interference filter is a glass or quartz plate C (Fig. 43), a few centimeters in diameter, well polished on both sides and coated with three successive layers: first, a semitransparent reflecting film M_1; next, a transparent coating D and, last, the second reflecting layer M_2. Another glass plate C then protects the coatings from mechanical damage. Because the incident rays reflect many times from the mirror surfaces M_1 and M_2 numerous interfering rays are created. This in turn results in an intensity distribution pattern in the transmitted light, whereby the passbands are sharply delimited.

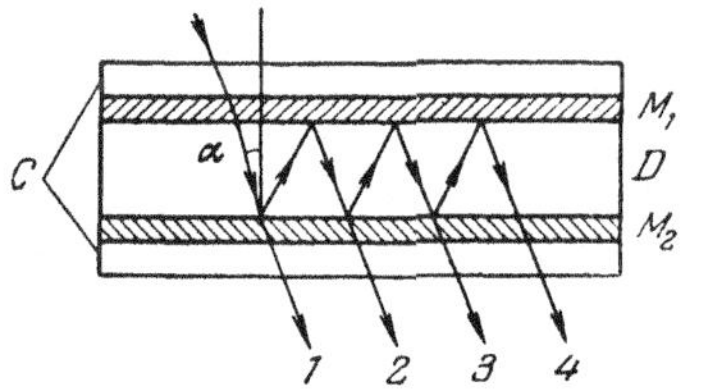

FIG. 43. The optical path in an interference filter.

Let n be the refractive index of the intermediate transparent layer D; l, its thickness; δ, the difference in optical paths between two successive rays 1 and 2; and α, the angle between the normal to layer D and the direction of the incident ray. The difference in optical paths between two successive rays 1 and 2 then becomes

$$\delta = 2nl \cos \alpha, \tag{3.34}$$

and the maximum intensity condition, which determines the transmission band of the filter, will have the form

$$2nl \cos \alpha = k\lambda_k. \tag{3.35}$$

That is to say, if the difference in optical paths between two successive rays is a multiple of the wavelength, the light filter will have a maximum transmission region. The greater the number of interfering rays, the narrower the transmission band.

The spectral half-width of the passband of a monochromatic filter is [327, 336]

$$\delta\lambda = \frac{\lambda^2}{2nl\cos\alpha}\,\frac{1-r}{2\pi\sqrt{r}}, \qquad (3.36)$$

where r is the reflection coefficient of layers M_1 and M_2.

It is evident from (3.36) that the monochromaticity of the transmitted radiation will increase with the thickness l of the transparent layer and the reflection coefficient r. However, increasing optical thickness of a light filter induces a complication, because the filter will eventually transmit a full spectrum of discrete monochromatic lines rather than a single band. If the optical thickness of the intermediate layer is equal to a half-wavelength of the visible light, then we have a *first order* filter with a single transmission band in the visible spectrum. If the intermediate layer is capable of accommodating k wavelengths of the visible light, then we have a filter of the kth order.

The factor $(1-r)/2\pi\sqrt{r}$ in (3.36) determines the number of interfering beams (N). The filter resolution, which is a function of N, may be as high as 20–50.

The reflectance and absorbance of the reflecting layers also determine the intensity of the light transmitted by the filter. The early types of interference filters used silver coatings as reflecting layers. Theoretical calculations [326, 327] have shown that interference filters with silver layers should transmit 45–50% of the visible light over a passband whose half-width is of the order of $\lambda/80$ (about 50–100 Å). However, the actual transmission curves for light filters proved to be far less satisfactory. The two shortcomings of this type of filter are the shift of the passband and a reduction of the peak transmission due to the oxidation of the silver.

In modern filters the silver films have been replaced by nonabsorbing multilayer dielectric coatings with high reflectance.

In this case the transmission factor may be as high as 80-90%, while the passband remains very narrow.

Methods for preparing dielectric coatings vary with different authors. Dufour [338], Polster [339], Korolev and Klement'yeva [340] obtained dielectric layers by vacuum deposition of zinc sulfide and cryolite. In these cases, the zinc sulfide also served as the intermediate layer, because this compound can give heavier coatings and thus produce higher order light filters. Krylova [341] used a chemical procedure to obtain dielectric films. She prepared multilayer coatings from alcoholic solutions of easily hydrolyzable ethyl esters of orthotitanic and orthosilicic acids, with subsequent heat treatment of the deposits. In her work, the number of deposited layers, averaging 1.2-1.4 microns in thickness, varied between three and fifteen.

Interference light filters with dielectric coatings can be prepared for the visible [336-342], UV [343, 344] and IR [342, 345] regions. Auxiliary tinted glass filters are used to eliminate secondary transmission peaks. However, these auxiliary filters increase the half-width of the passband and reduce the transmission factor.

Table 1 summarizes data of some selected authors [336, 342, 343] on the characteristics of various interference filters with multilayer dielectric coatings.

Monochromatic interference filters obtained by vapor deposition of zinc sulfide and cryolite have transmission bands with smaller half-widths and considerably narrower curve tails than those prepared chemically from TiO_2 and SiO_2.

The half-width of the transmission band can be further reduced by combining several interference filters. Half-width values of the order of 1 Å or less at a transmission factor of about 70% can be obtained with these composite (or multiplex) interference filters [336, 346].

Table 1

Characteristics of interference filters

λ, millimicrons	T_m, %		2δλ, millimicrons		Refs.
	without auxiliary filters	with auxiliary filters	without auxiliary filters	with auxiliary filters	
850	80	-	35	-	342
700	80	56	28	23	342
652	65	-	3.57	-	336
589	70	50	-	12	342
549	60	45	-	11	342
477	70	54	-	10	342
440	72	47	-	10	342
405	60	43	-	6-8	342
386	50	35	-	6	342
390	82	40	14	16	343
330	83	43	14	10	343
310	76	38	12	6	343
290	76	33	6-8	8	343
270	65	27	6	8	343
250	53	-	8	-	343

The theory and the procedure for construction of a multiplex interference light filter are discussed by Korolev [336]. He prepared a complex light filter with the following parameters: $\lambda_m \approx 540$ millimicrons; $2\delta\lambda = 3.3$ Å; $T_m \approx 50\%$. From this, one can conclude that interference filters achieve about the same monochromatization as the conventional monochromator. Unlike the latter, however, an interference filter can have a high transmittance (aperture ratio), which greatly simplifies the photoelectric detection of weak luminous fluxes. This is a great advantage in some special problems of spectral analysis of gases.

The tabulated filter data give λ_m values for the case when the direction of the incident light coincides with the normal to the filter. If the light falls obliquely, the position of the passband peak shifts somewhat. Such shifts are sometimes useful because they permit some minor variations in the passband. However, obliquely incident light reduces the aperture ratio of the filter and slightly increases the value of $\delta\lambda$.

The passband of an interference filter is rigorously fixed. Thus analytical work involving a variety of spectral ranges requires a set of filters [347]. To avoid dealing with many filters, one can construct a variable-thickness interference filter, in which case the transmitted wavelength can be altered by selecting some fractions of the filter. Such filtering devices are known as optical wedge interference filters [318]. This type of filter cannot be designed for a broad spectral range, since the small surface area of the filter makes it difficult to obtain a steep transmission curve (large shifts of the passband with small changes in filter thickness) along with a large emerging light flux.

Unfortunately, good interference filters are difficult to construct. Broad-band filters, on the other hand, lower the sensitivity of the analysis (see Section 26), since presence of transmission band tails is equivalent to the presence of background noise.

14. PHOTOELECTRIC SPECTROMETERS

Photoelectric techniques are now finding increasing use in spectral analysis, replacing the earlier photographic procedures. Photoelectric systems are highly accurate, have a very fast response, and in many cases can be completely automated.

The photographic methods may still be used to advantage in qualitative analysis, since spectrum photographs usually are more familiar and therefore easier to interpret than a graph produced by a recorder. In addition, in some cases, it may be possible to obtain photographs in the same time it would take to obtain a recording via photoelectric means, and the photograph may also give more information. This is because a photographic procedure permits the simultaneous recording of a virtually unlimited number of elements of the spectrum, while even the most efficient photoelectric detectors are incapable of responding to more than a few dozen such elements.

Photoelectric systems usually consist of a light source, the spectral instrument, a radiation detector with a power supply unit, and a recorder.

The spectral instrument may be a prism or grating spectrograph, with a set of fixed slits for separating the desired wavelengths provided in the cassette compartment. Monochromators and monochromatic light filters may also be included. In many cases, the use of filters considerably simplifies the remainder of the system.

Photocells or photomultipliers coupled with a power supply are used as radiation detectors. Depending on the intensity of the light flux measured, DC or AC amplifiers may or may not be needed. An AC amplifier offers certain advantages, inasmuch as it has no zero drift, does not require a highly stabilized power supply, and eliminates the need for dark current compensation by means of auxiliary components.

We shall not dwell here on the currently available radiation detectors which have already been described in sufficient detail by Chechik et al. [348]. We shall consider here only a few typical systems which are currently used or may eventually find application in the spectral analysis of gases.

The photometer assembly proper consists of radiation detector with a power unit, a receiver-amplifier circuit and a recording system. Depending on the method of measurement modern photometers are either direct- or zero-reading. All methods of quantitative spectral analysis of gas mixtures measure the relative intensities of the lines of the desired component and reference substance. For this reason, both direct- and zero-reading methods are used in the photoelectric technique.

Direct reading photometers

In this type of photometer the measured lines are projected one after another onto the photocathode of the photomultiplier. Alternatively, one can use as many radiation detectors as there are lines to be measured. In gas analysis, one normally measures five components at the most. The spectrum is scanned so rapidly that even single-channel direct-reading systems can be used effectively. Multichannel installations usually require greater attention and involve difficulties (e.g., matching the photomultiplier characteristics and compensating for the drift of the electrical gear).

Let us now consider a few typical direct-reading photometers.

The photoelectric system of the Physics Institute of the Leningrad State University is a self-contained plug-in unit consisting of a photomultiplier with a voltage divider enclosed in a separate housing, a DC amplifier with a powerpack, and a stabilized rectifier to supply the photomultiplier. The design is based on the Rosenberg bridge circuit, adapted by Osherovich et al. [349]. A schematic of the photometer is given in Fig. 44 and its photograph in Fig. 45. A single-stage DC amplifier tube A is connected to one arm of the bridge. The resistor R_{grid} in the circuit of the last photomultiplier stage provides the input resistance for the amplifier. An additional potentiometer-regulated bias compensates for the direct-current component of the photomultiplier dark current. The circuit parameters are selected in such a way that it is possible to adjust the galvanometer scale indicator to zero at a small grid current, both when the discharge tube is switched on and when it is dimmed. The photocurrent may be read on a microammeter with sensitivity of 10^{-7} A/scale division. Alternatively, one may use a recording potentiometer with a maximum scale reading of 10 mV and 200 Ω shunting resistor across the input.

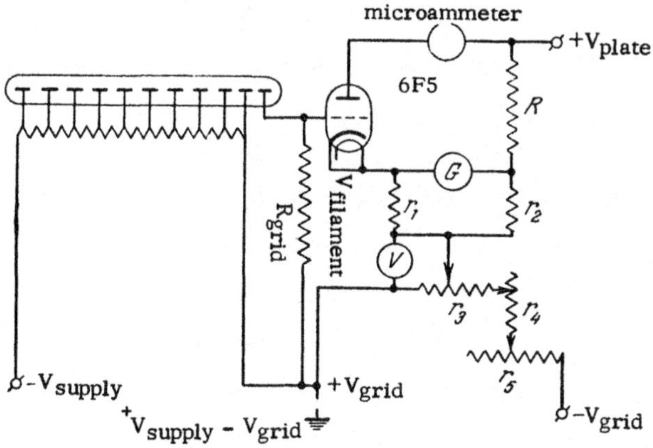

FIG. 44. The photometer of the Physics Institute of the Leningrad State University (with DC amplifier). R_{grid} = 11 mΩ; $r_1 = r_2$ = 3 kΩ; $r_3 = r_4$ = 10 kΩ; r_5 = 500 Ω; R = 0.5 MΩ; V_s = 1100 V; V_p = 160 V.

This photometer module, combined with an ISP-51 spectograph (see Appendix III) or a monochromator, can be used for most spectroscopic gas analyses (neon-helium mixture in air, helium in neon, neon in helium, nitrogen in argon, nitrogen and oxygen in helium). When coupled with a grating spectrograph such as the DFS-3 (see Appendix III), this photometer may also be used for determining the isotopic composition of hydrogen. The DC amplifier

FIG. 45. The photometer of the Physics Institute of the Leningrad State University external view.

may be omitted if the spectrum of the desired range is isolated by means of interference filters. The intrinsic error of the system is negligible if a stable power supply is used.

Photometer modules PS-381, PS-382 and FEP-1, described on the following pages, are designed to work with the ISP-51 spectograph. They incorporate recorders and their operation is based on scanning of the spectrum at the exit slit of the spectrograph. Thus, the light falling on the entrance slit of the spectrograph is dispersed into a spectrum. With the aid of a rotating prism assembly, various portions of the spectrum are projected successively through the exit slit upon the photocathode of the photomultiplier. The signal thus generated is amplified in a DC amplifier and is fed to the recorder. The recorder chart drive is synchronized with the prism-rotating mechanism. The spectrum is recorded as a continuous curve, on which time markers appear at equal intervals. As the prism system revolves, the focusing lens is automatically shifted, thus continuously focusing the spectrum on the exit slit.

The receiver-amplifier circuit of the *PS-381 instrument* [350] is shown in Fig. 46. The light is detected by photomultiplier FEU-17

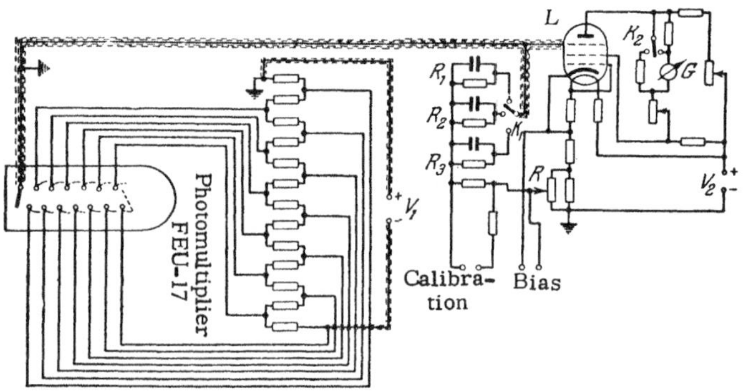

FIG. 46. Receiver-amplifier circuit of the PS-381 photometer module. R_1—1 MΩ; R_2—10 MΩ; R_3—100 MΩ; K_1, K_2—switches; L—amplifier tube; G—galvanometer; R—potentiometer; V_1 = 750-1000 V; V_2 = 12-15 V.

(see Appendix IV) and amplified by a DC amplifier tube connected in a bridge-type circuit. The 1, 10 and 100 MΩ resistors in the control grid circuit permit varying the gain by a factor of 1, 10 and 100, respectively. With the 100 MΩ resistor, the gain $K = 10^4$. Potentiometer R compensates for the direct current component of the dark current. The photocurrent is indicated by means of a sensitive mirror-type galvanometer and a recording of it is obtained by projecting the light beam reflected by the mirror upon a strip of photosensitive paper. This photometer (see Fig. 47, a) is designed for the range of 4000-5400 Å; focal length of its exit collimator lens is 300 mm, its effective aperture ratio is 1:6 and its recorder provides a reproducibility of 5%.

Photometer PS-382 [351] is designed for the range of 3600 to 9500 Å. Two photomultipliers, FEU-17 and FEU-22 (see Appendix IV), function as radiation detectors. Both units are stationary and work in sequence, whereby the supply voltage is switched from one to the other. Otherwise, the circuit is the same as in the PS-381 instrument. The focal length of the exit collimator is 800 mm; its aperture ratio is 1:16. The linear dispersion of the prism is 11 Å/mm at 4400 Å and 50 Å/mm at 7000 Å.

Photometer FEP-1 (see Fig. 48 and [352]) is designed for the range of 3800-6600 Å; the focal length of its exit collimator lens is 300 mm; the aperture ratio is 1:6. The detecting-recording system is of improved design. The light is detected by photomultiplier FEU-17 to which power is supplied from a voltage divider connected to a stabilized high voltage rectifier. The high voltage stabilizer is supplied from a rectifier made up of vacuum tube diodes a connected in a half-wave rectification circuit. The detector tube b is a triode and the amplifier c is a pentode. The screen grid reference voltage is obtained from voltage regulators d and e. The positive stabilized voltage terminal is grounded.

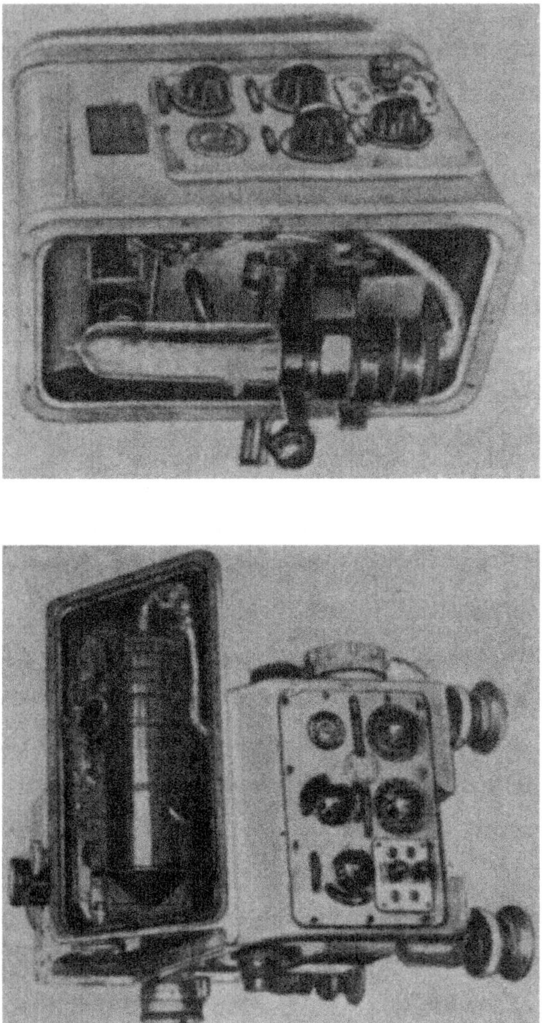

FIG. 47. Exit collimators of the PS–381 (a) and PS–382 (b) instruments with open photomultiplier chambers.

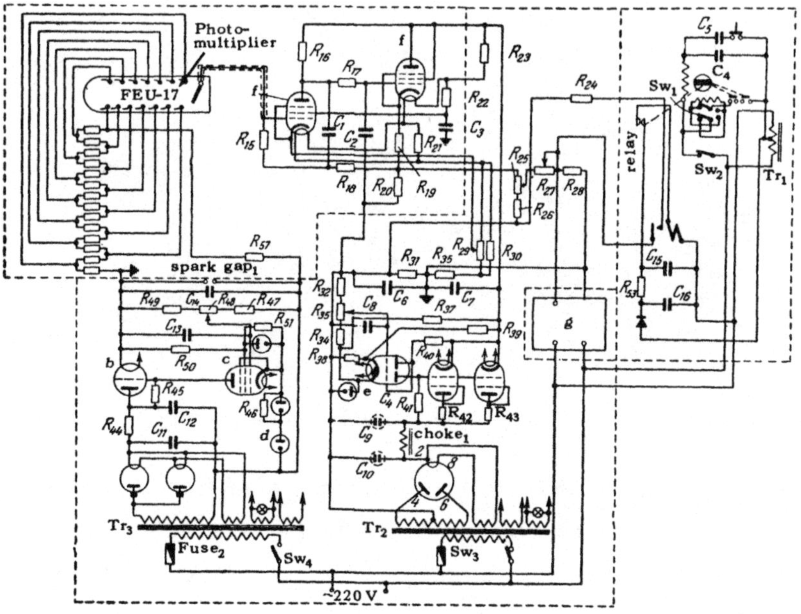

FIG. 48. Receiver-amplifier circuit of the FEP-1 instrument.

The signal from the photomultiplier is amplified by a two-stage DC amplifier circuit consisting of tubes *f*. The small filament current (about 75 mA) can be supplied by an electronic voltage stabilizer.

The input amplifier *f* operates in a conventional pentode hookup, but with lower electrode voltages (plate, 30 V; screen grid, 20 V; filament, 10.4 V). This amplifier gain is 600–800. The second tube *f*, connected as a triode, operates as a cathode follower, with a fraction of the load returned as the feedback voltage to the amplifier input. The signal is recorded by a compact recording potentiometer *g*. The recording system is a self-contained unit, with the amplifier and power supply components mounted in the lower and the recording potentiometer in the upper compartment. A jacketed coaxial cable connects the recorder to the photomodule. A general view of the FEP-1 instrument is shown in Fig. 49. Both the

FEP-1 and the other above described photometers have been used effectively for analysis of neon, nitrogen and hydrogen in helium.

FIG. 49. General view of the FEP-1 instrument.

The time required for spectroscopic analyses of gas mixtures can be considerably reduced by using equipment with direct photoelectric readout. However, the accuracy of the analysis suffers because the intensities of lines being compared are measured in succession. Such procedures, furthermore, fail to eliminate the analytical error due to unstable excitation conditions.

A design incorporating a direct readout and using an AC amplifier, proposed by Abramson [353], obviates some of these problems. In this arrangement, two factors counterbalance the effect of unstable excitation conditions. For one thing, the system integrates the measurement [354]. Secondly, line intensities obtained for a given element are compared with the intensity of undispersed light from the discharge. A block diagram of the Abramson unit is shown in Fig. 50. Its operating principle consists of the following:

The light emerging from the source P falls on the entrance slit S_1 of the ISP-51 spectrograph (see Appendix III), in which the camera is replaced by a collimating lens with a focal length of 50 cm. The spectral line of the element undergoing determination is separated as the light passes through exit slit S_2. The transition from one line to another is achieved by means of a rotating prism system. Radiation detector F_1 receives the light of the isolated spectral line. Instead of a reference line, the comparative evaluation is based on the total radiation reflected from the frontal facet of the first prism, and received by the radiation detector F_2. The output signals from detectors F_1 and F_2, amplified respectively by AC amplifiers A_1 and A_2 and rectified by solid state devices D_1 and D_2, are fed to condensers C_1 and C_2 through identical resistors R_1 and R_2, thus building up the condenser charges. At the end of the analysis, the potentials on the charged condensers are proportional to the mean values of the light fluxes received by the radiation detectors. The concentration of the desired component can then be determined from the ratio of the condenser voltages.

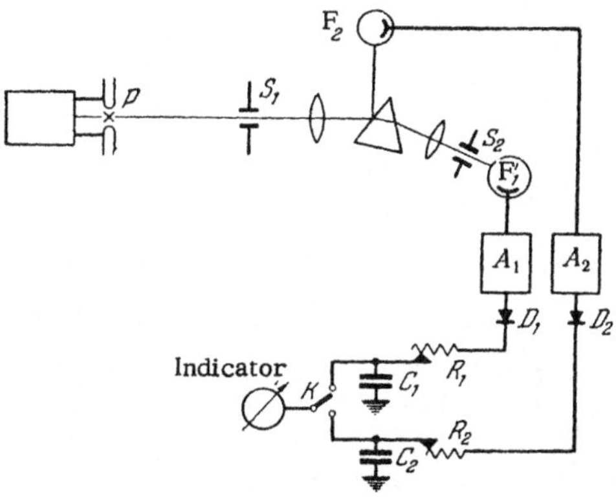

FIG. 50. Block diagram of the Abramson's apparatus.

There is an alternative method for determining concentrations. During charging of the condensers a voltmeter is first connected through a two-way switch K to condenser C_1. When the potential on this condenser reaches a certain value, K connects the voltmeter to condenser C_2 and at the same time shuts off the current to the discharge tube. The system can be precalibrated for direct measurement of concentrations in the mixture.

Some features of this unit must be modified in adapting it to meet the needs of practical gas spectroscopy. For one thing, the total radiation emitted by a discharge tube depends a great deal on the composition of the gas mixture, and hence it cannot always be used for direct comparison. The emission due to the principal component of the sample can be separated from the total radiation by means of a filter. In addition the amplifier circuit must be chosen in accordance with the type of light source used. If the source is a DC discharge tube, the light must be converted to an AC signal. On the other hand, if one works with high frequency discharges, the oscillator frequency should not exceed the limit of the amplifier passband.

Null type photometers

Null type photometers [354-359], now widely used in spectroscopic analysis, fall into three groups:

a) photometers designed for simultaneous comparison of light fluxes (compensation method);

b) photometers in which the measured photocurrent is compared with a current from an outside source (substitution method);

c) photometers whose operation is based on alternate comparison of two measured photocurrents (the method of flickering).

In the *compensation procedure* the simultaneous comparison of measured photocurrents requires the use of not less than two radiation detectors (one for the reference line and the other for the line of the desired component of the sample). The photocurrents, which correspond to the respective luminous fluxes, are switched on in opposition to each other. One of the measured signals is attenuated, either electrically, or by means of an optical wedge. Electrical attenuation is achieved either by varying the voltage of the photomultiplier supply, or by inserting a calibrated voltage divider in the control grid circuit of one of the amplifying stages. The indicator is adjusted until it reads zero (both photocurrents are equal). Then the ratio of light fluxes is equal to the attenuation factor of the stronger current. The accuracy of the value thus determined depends entirely on the accuracy of calibration of the optical wedge or of the voltage divider.

Some of the devices which simultaneously measure two light fluxes operate with modulated photocurrents whose respective frequencies f_1 and f_2 are not multiples of each other. Both light fluxes are directed onto the same photomultiplier with the photocurrents then amplified by two identical amplifiers tuned, respectively, to the frequencies f_1 and f_2. The rectified photocurrents are compared following the detection [360]. This type of photometer uses a two-channel recorder, and careful control must therefore be exercised over both channels (both must have identical gain).

The *method of flickering* [348] is based on the alternate measurement of the two luminous fluxes, which are compared by means of a single photomultiplier. Both fluxes are directed to the same fraction of surface of the photocathode, and each is alternately shut off by a chopper. If the chopping frequency is sufficiently high and the two fluxes differ in intensity, a pulsating direct current will flow through the photomultiplier plate circuit. This current can be read out by means of a filament voltmeter or "magic eye" tube. As one of the fluxes becomes less intense, the pulsating voltage amplitude decreases and approaches zero when the two fluxes are equal in intensity.

The advantage of photometers based on the zero-reading principle is that the two values undergoing comparison are determined simultaneously. Furthermore, working with frequency-modulated fluxes and a single photomultiplier eliminates additional measurement errors resulting from the use of two or several photomultipliers.

The simplest arrangement for measuring the ratio of two values (Fig. 51) by the compensation method is that presented by the Soviet gas analyzer SF-4101 [358] which is designed for nitrogen determination in argon (see Section 26). The light fluxes undergoing comparison are separated by monochromatic filters, so that vacuum-tube photocells can be used as radiation detectors. A cathode follower and an electronic recording potentiometer are used to measure and record the relative strengths of the photocurrents.

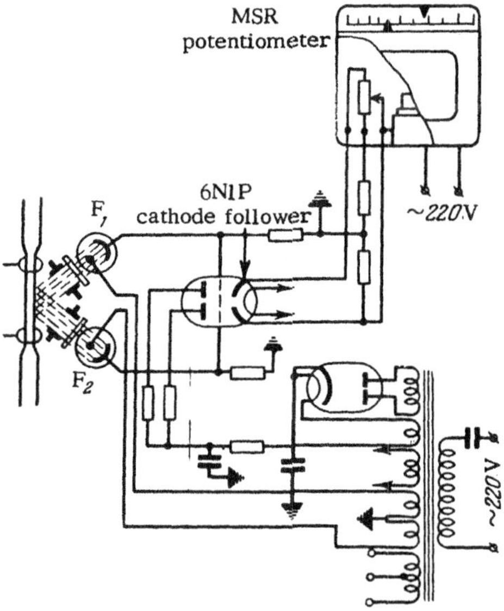

FIG. 51. Circuit arrangement of the SF-4101 gas analyzer for simultaneous recording of two light fluxes.

A similar circuit arrangement (Fig. 52) for indicating the ratio of intensities of two luminous fluxes is used in the gas analyzer for nitrogen determination designed at the R & D instrument shops of the Physics Research Institute of the Leningrad University (see Section 26). Two FEU-19 photomultipliers (see Appendix III) function as radiation detectors. The photomultiplier signals are fed without preamplification to the input of a potentiometer whose slide-wire acts as the balancing element. The disadvantage of this arrangement is that the measured data correspond to instaneous light flux values and there is no time-averaging (integration) of the results. This considerably reduces the reproducibility of the measurements. This shortcoming can be eliminated by using circuits with charge accumulation on a condenser [354].

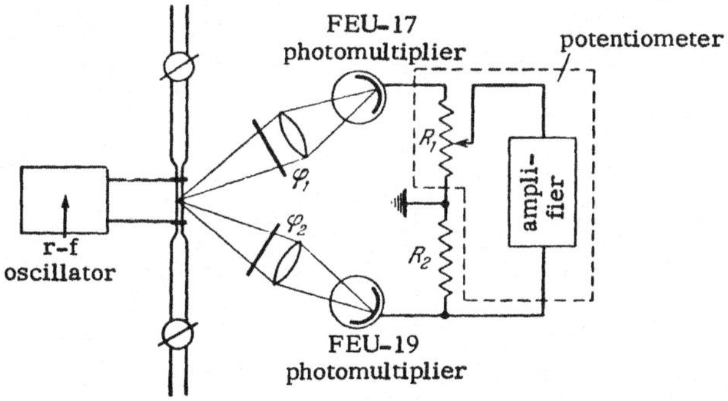

FIG. 52. Diagram of the experimental gas analyzer of the Physics Research Institution of Leningrad University. Simultaneous recording of two light fluxes. φ_1, φ_2—filters; R_1—slide-wire resistance; R_2—variable resistor.

The charge accumulation method of measuring relationships between light flux intensities was used by Lee [361] for argon determination in nitrogen (Fig. 53). As the first step in measuring the intensity ratio, integrating condensers C_1 and C_2 are charged for 30 seconds with photocurrents from photomultipliers F_1 and F_2 respectively. By the end of this charging period, the ratio of condenser potentials is proportional to the photocurrent ratio and, therefore, to the argon concentration in nitrogen. After charging, switches K_1 and K_2 close and the condensers discharge across identical resistors R_1 and R_2. Relay switches P_1 and P_2 connected to the plate circuits of the two DC output amplifiers are triggered simultaneously with the closing of the switches K_1 and K_2. The recorder circuit is now open. As the condensers keep discharging, the output current diminishes. The charge accumulated on condenser C_2 is smaller than that on condenser C_1, since the impurity (argon) line is less intense than the reference line (a requisite condition of the operation of the circuit). As a result, the potential across condenser C_2 and hence the output current flowing through relay switch P_2 will eventually reach the value at which the relay is set into operation closing the recorder circuit. A line is traced on the recording drum as long as the upper relay switch (P_1) remains closed. When the potential across the discharging condenser C_1 has diminished to a point where it is equal to the potential on condenser C_2, then the relay switch P_1 opens and the tracing on the drum stops. The length of the tracing will be proportional to the condenser discharge time differential $t_1 - t_2$, and will thus be proportional to the logarithm of the intensity ratio of the two spectral lines compared.

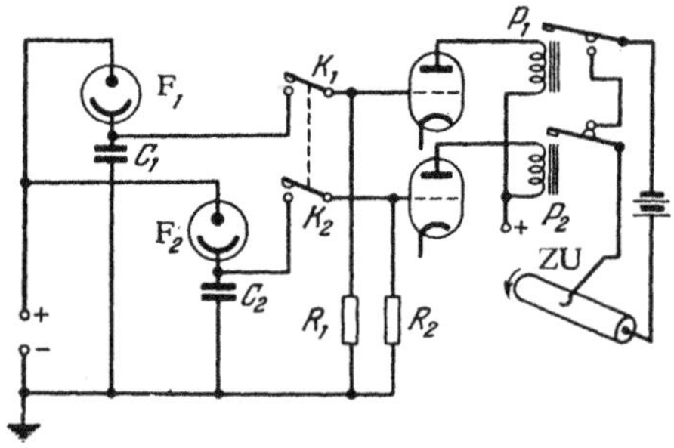

FIG. 53. Charge-accumulation circuit for simultaneous measurement of two light fluxes. ZU—recording drum; P_1, P_2—relay switches.

Lee used the same procedure to determine argon in nitrogen on the basis of absolute intensities of the argon lines. The second condenser was charged by current supplied from a steady outside source; i.e., the substitution method was used.

The common drawbacks of all multichannel systems for simultaneous measurement of two light flux intensities are completely eliminated in circuits which use a single radiation detector [348].

Figure 54 demonstrates a circuit arrangement in which the intensities of two fluxes are compared with a single photomultiplier (method of flickering). The system was used in determinations of nitrogen in argon [359]. The two fluxes are projected alternately upon the photomultiplier by means of a revolving slotted chopper disk. The interruption frequency is 20 cps. The photocurrents are amplified (50-fold) by a transistorized emitter-follower and are then indicated on a ratiometer, whose terminals are alternately connected to the emitter by a synchronous relay switch.

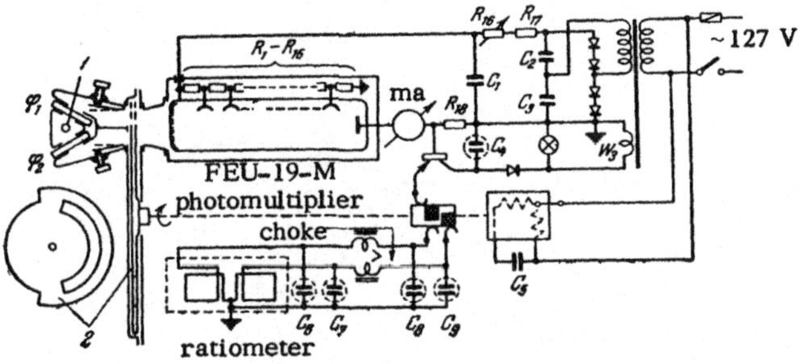

FIG. 54. Successive detection of two fluxes by a simple photomultiplier. 1—Discharge tube; 2—chopper disk; φ_1—filter with λ_m = 390 millimicrons; φ_2—filter; R_1, R_2 ... R_{15} = 15 kΩ; C_1, C_2, C_3 = 1 μF; C_4 = 1000 μF; C_5 = 2 μF; C_6, C_7, C_8, C_9 = 1000 μF; R_{17} = 43 kΩ; R_{16}, R_{18} = 0.1 MΩ.

The drawback of this system lies in the alternate indication of the two fluxes. Figure 55 shows the diagram of a photometer with a single radiation detector [362], designed for simultaneous measuring of two fluxes and recording of their intensity ratio. The frequency-modulated fluxes (f_1 = 930 cps; f_2 = 2000 cps) are directed onto the same area of the multiplier photocathode, so that a composite output signal is produced from the two sets of oscillations. The composite signal is amplified, so that the ratio of amplitudes

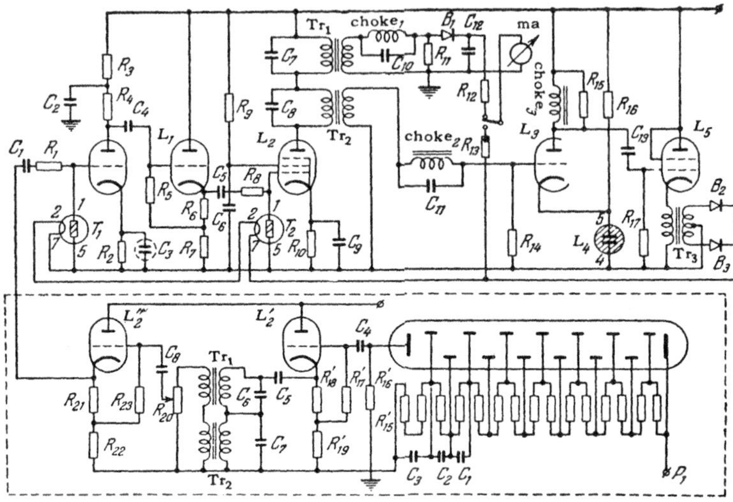

FIG. 55. Single photomultiplier circuit for recording the intensity ratio of two frequency-modulated fluxes. R_1, R_8 = 10kΩ; R_2, R_6 = 2kΩ; R_3 = 5.1kΩ; R_4 = 51kΩ; R_5, R_{17}, R_{22} = 510kΩ; R_7, R_9, R_{12} = 100kΩ; R_{10} = 200Ω; R_{11}, R_{14} = 1MΩ; R_{15} = 220 kΩ; R_{13} = 12-18 kΩ; R_{16} = 15 kΩ; R_{18} = 4.7 MΩ; R_{19} = 2 MΩ; R'_{16} = 1 MΩ; R'_{17} = 2 MΩ; R'_{18} = 1 kΩ; R'_{19} = 51 kΩ; R_{20} = 220 kΩ; R_{21} = 2 kΩ; R_{22} = 82 kΩ; R_{23} = 510 kΩ; C_1 = 0.07 μF; C_2 = 10 μF; C_3, C_9, C_{17} = 30 μF; C_4, C_5 = 0.05 μF; C_6 = 0.1 μF; C_7 = 0.01-0.05 μF; C_8 = 0.01-0.1 μF; C_{10} = 0.01-0.05 μF; C_{11} = 0.01-0.1 μF; C_{12} = 50 μF; C_{13} = 0.25 μF; C_{14} = 0.03 μF; C_{15}, C_{16} = 0.1 μF; C_{18}, C_{20} = 20 μF; C_{19} = 0.25 μF; C_{21}, C_{22} = 500 μF; L'_2, L''_2, L_1, L_3—Soviet tube 6N1P; L_2—selective amplifier; L_4—voltage regulator; L_5—detector tube; T_1, T_2—thermistor; B_1, B_2, B_3—diodes.

remains unaltered. Following amplification, the combined signals are separated by means of a selective amplifier tube which has two tank circuits connected to the plate circuit. The f_2 signal is now fed directly to the measuring instrument while the f_1 signal is delivered to the so-called subtracting stage, to which a fixed voltage from stabilitron tubes is fed as a reference voltage. The amplified voltage differential $V_{f_1} - V_{\text{stabilitron}}$ is used for the negative feedback—the latter designed to keep constant the output voltage V_{f_1} of the selective amplifier. Two thermistors with indirect heating and a ballast resistor function as control elements of the electronic circuit. Because of the presence of two control elements in the measuring circuit, accurate intensity ratio data can be obtained for a widely variable (by a factor of 15-20) output signal level. In theory, the circuit can be adapted for measuring more than two values. The usefulness of this arrangement for gas analysis was tested in nitrogen determinations in argon. The light fluxes being compared were separated by means of monochromatic filters. The tests yielded good results.

None of the above described systems of photoelectric indication is ideal or applicable to all cases.

In analytical procedures with a photoelectric indication the sensitivity limit is determined by the effect of the continuous background and by the dark photocathode current. The latter effect can be eliminated either via compensation or by use of AC amplifiers.

Gas analyzers

We shall now describe a few spectroscopic gas analyzers.

The automatic gas analyzer SF-4101 [358] was developed by the Soviet State Bureau for the Design and Development of Analytical Instruments specifically for nitrogen determinations in argon in the concentration range of 0 to 1%. The diagram of this analyzer is shown in Fig. 56. A forepump causes a continuous stream of gas to flow through a 5 mm I.D. discharge tube at a pressure of 0.3 mm Hg. The pressure is automatically regulated by means of control valves *2* and *4*. Tube *A* hooked up in the bridge circuit of an amplifier *D* operates as a pressure transducer. The needle of the control valve is driven by a servo-motor *B*. An r-f, 30 W, 20 Mc push-pull oscillator *C*, powered by a separate 600 V rectifier, is used for exciting the discharge. The luminous fluxes emitted by nitrogen and argon are separated by filters.

The photoelectric recording circuit was described earlier. The analyzer scale is calibrated directly in percent nitrogen. The readout time-lag with respect to the change in nitrogen concentrations at the analyzer input depends on the travel speed of the carriage of the electronic potentiometer. If a drying filter is inserted in the path of a gas stream flowing at the rate of 0.5 liters/min, the time lag will be as high as 15-30 sec. The readings are reproducible within 5%.

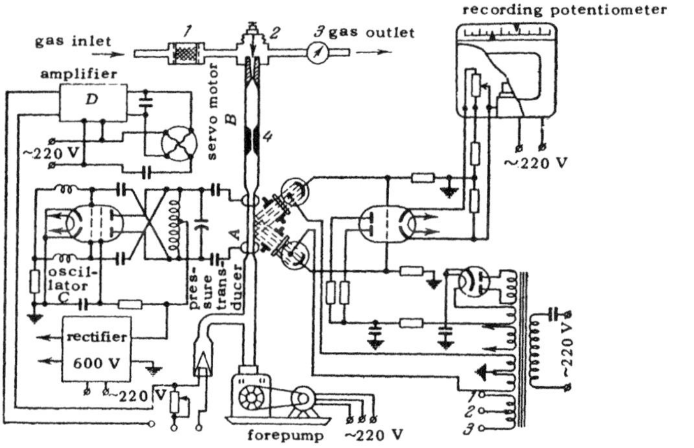

FIG. 56. Schematic diagram of the SF-4101 gas analyzer.

By varying some of the circuit parameters the equipment can be adapted for determining nitrogen concentrations ranging down from 0.2%. The analyzer has several discharge tubes which are interchangeable without disturbing the vacuum existing in the unit. The recorder is mounted in the upper part of the module; the center part (which is the gas analyzer proper) consists of a discharge tube and a receiver-amplifier unit, each with its own powerpack, and the lower part houses a forepump. A general view of this gas analyzer is shown on the right side of Fig. 57.

The *gas analyzer* of the Physics Institute of Leningrad University [418-420] is designed for nitrogen determinations in argon and helium, at concentrations in the range of 0.001-1%. The analyzer

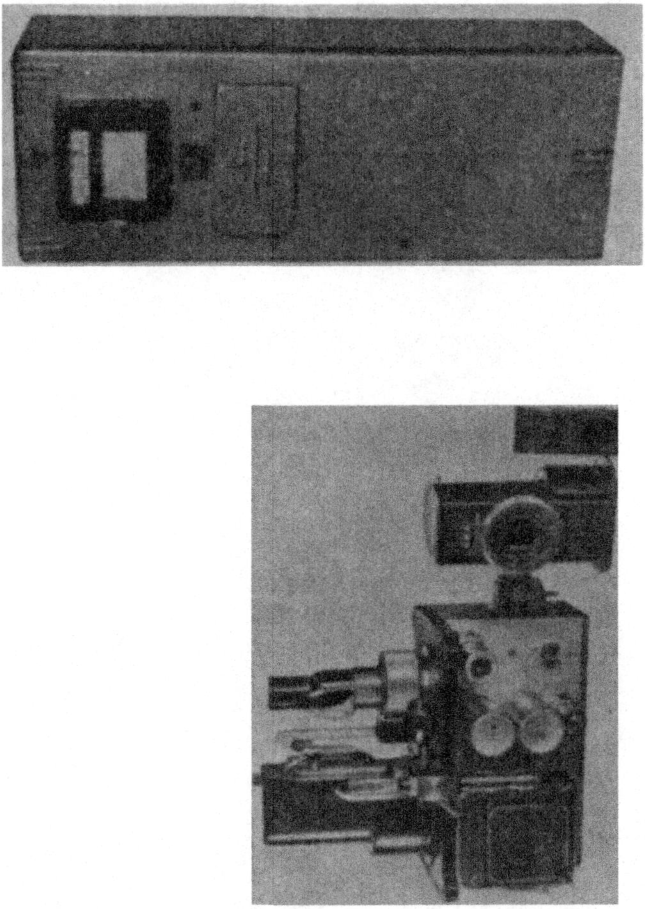

FIG. 57. Outside view of the SF-4101 gas analyzer and of the internal unit with discharge tube.

is a self-contained apparatus (see Fig. 58) and consists of standard modules. These include two photomultipliers FEU-19 (see Appendix III) with voltage dividers, mounted in separate housings. The focusing lenses and the photomultiplier power supply filter are mounted on the same cases. The other units are a rectifier, an r-f oscillator with a power supply unit, and an indicating or a recording potentiometer.

FIG. 58. The general view of the gas analyzer of the Physics Research Institue of Leningrad University.

The vacuum section of the analyzer (see Fig. 59) consists of a discharge tube *1*, a *U*-tube manometer *2* which measures the pressure in the tube, and stopcocks *3*, *4* and *5* which control the continuous gas flow. Stopcock *5* is in the gas cylinder line and stopcock *4*, in the forepump line. Pressure control is achieved by means of the same stopcocks *4* and *5*. Calibration curves plotted for standard mixtures are furnished with the analyzer. The calibration of the analyzer must be rechecked periodically, as well as in cases where the discharge tube has been replaced. The reproducibility of data is about 5%. When working with a set of interchangeable interference filters, the analyzer can determine hydrogen in helium in the range of 0.001 to 1% and higher; neon in helium, in concentrations ranging from 0.001% and up; hydrocarbons in helium, in the 0.001-0.1% concentration range; carbon dioxide in argon, at concentrations of 0.005 to 0.05%.

FIG. 59. The vacuum module of the gas analyzer of the Physics Institute of Leningrad University.

Other photoelectric units are described in review [363]. The application of photomultipliers and of electronic devices in spectroscopic determinations is described in [348] and [364], respectively.

Chapter IV

METHODOLOGICAL PROBLEMS

15. RELATIONSHIP BETWEEN LINE INTENSITY AND CONCENTRATION OF AN ELEMENT IN A MIXTURE

The objective of quantitative spectral analysis is to establish a relationship between line intensity and concentration of an element in a mixture. Existing analytical procedures are based for the most part on determination of the relative intensities of the lines of the element being analyzed and those of a comparison (reference) element [25, 315]. The comparison standard is either the principal constituent of the mixture or an element added for this purpose.

Two cases should be distinguished: when the ionization potential of the admixture (or minor constituent) to be determined is greater than that of the principal component, and vice versa.

Let the principal component be denoted by A, the minor one by B, and their ionization potentials by V_A and V_B, respectively.

Consider the first case: $V_B > V_A$. If the concentration of B is low, its changes will not affect the discharge conditions. Consequently, $f(T_e)$ = const and n_e = const. Then from relationship

(1.7) we can write for the intensities of spectral lines A and B

$$I_A = aN_A, \quad I_B = bN_B, \tag{4.1}$$

where a and b are proportionality constant, while N_A and N_B are the concentrations of normal atoms of the components in the mixture. The value of N_A can be considered constant, while N_B is proportional to the concentration of element B in the mixture (C_B). Then

$$\frac{I_B}{I_A} = \alpha C_B, \tag{4.2}$$

where α is a proportionality factor. Therefore, in this case the ratio of line intensities of the minor to the principal constituent varies linearly with the concentration of the minor constituent.

Next, consider the second case: $V_B < V_A$. Then the electron temperature cannot be assumed constant, since the concentration of the readily excited component varies (see Section 2). Hence

$$I_B = N_B f_B(T_e), \quad I_A = N_A f_A(T_e), \tag{4.3}$$

where $f_B(T_e)$ and $f_A(T_e)$ are monotonically increasing functions of the electron temperature.

The electron temperature falls with a rise in concentration of the readily excited component. This causes an overall decrease in line intensities which is more pronounced in the lines of the difficult-to-excite component. Thus, a higher ratio of line intensities of the minor and the primary constituent is due to two factors: a lower electron temperature and a higher concentration of the minor constituent. It follows that

$$\frac{I_B}{I_A} = \alpha f(N_B) N_B, \tag{4.4}$$

where $f(N_B)$ is a montonically increasing function of the concentration of the atoms of the minor constituent (N_B). Expressing N_B in terms of C_B., we have

$$\frac{I_B}{I_A} = \beta f(C_B) C_B. \tag{4.5}$$

Here β is a proportionality factor, while $f(C_B)$ is a montonically increasing function of the concentration of B.

Thus the relative intensity of the lines of the minor and principal substances is no longer a linear function of the concentration of the minor constituent.

In isothermal sources the temperature, as a rule, does not vary appreciably with the concentration of the minor constituent. Hence there is a linear relationship between the intensity of the line for that constituent and its concentration in the discharge, provided no reabsorption is involved.

In order to make a judicious choice of operating conditions for a gas mixture analysis it is essential to determine the effect of the various discharge parameters on the ratio of line intensities of the two components in the gas mixture. Assuming one of the components to be easy to excite and the other difficult to excite, we shall first show how the above ratio varies with pressure, current and tube diameter.

The electron temperature decreases with pressure, which in turn decreases the line intensities of both components. If $V_A > V_B$, a decreasing temperature will cause a sharper decrease in the intensity of the line for A than in that for B. Hence the ratio I_A/I_B will decrease with increasing pressure. It follows that a high pressure is necessary for determining a small quantity of a readily excited impurity in a difficult-to-excite primary substance. Analogous considerations lead to the conclusion that a low pressure must be used for determining a small admixture of a difficult-to-excite component in a readily excited primary substance. Actually, some optimal intermediate pressure is usually chosen; it should not be too low to avoid an excessive reduction of the total radiation intensity.

The electron temperature also drops with an increase in the discharge tube diameter [91, 99]. The sensitivity of spectral analysis for difficult-to-excite elements will thus improve in narrow discharge tubes. The effect of the discharge tube diameter on the relative line intensities of helium (λ = 5876 Å) and argon (λ = 6416 Å) is shown in Fig. 60 [365]. For low currents the ratio of line intensities increases with the current up to saturation. In some cases a further increase in the current may reduce the line intensity ratio.

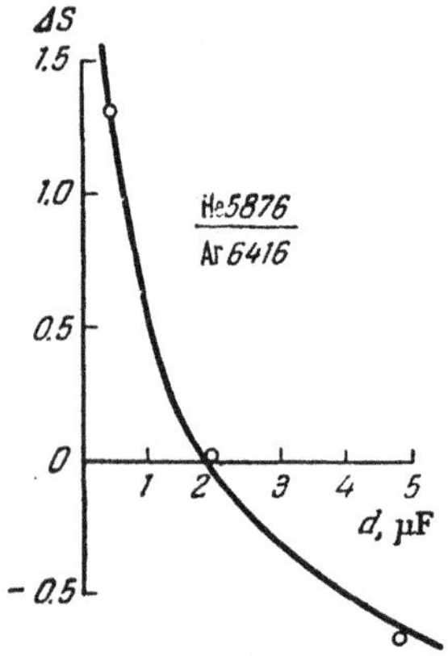

FIG. 60. Effect of the diameter of the discharge tube on relative line intensities of helium and argon.

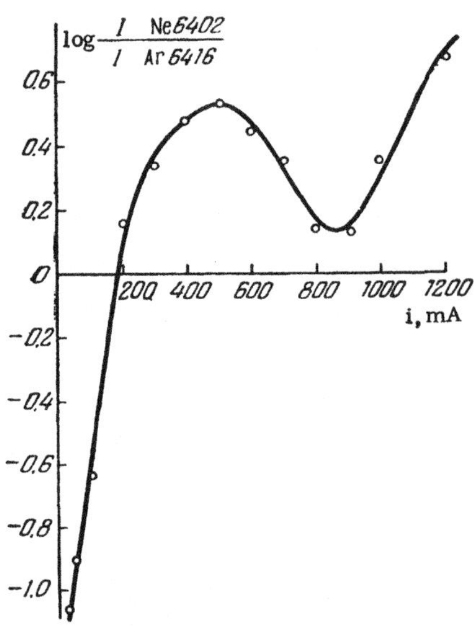

FIG. 61. Effect of the discharge current on the relative line intensities for neon and argon.

The effect of the current on the relative line intensities for neon (λ = 6402 Å) and argon (λ = 6416 Å) is shown in Fig. 61. The conditions were: tube diameter, 20 mm; tube pressure, 0.7 mm Hg; 10% argon in neon. The observed lack of monotonicity in the functional relationship is readily understood if we recall that even the absolute intensity of a spectral line may decrease with increasing current [94]. The intensity decline varies in different lines, because the line intensity depends on both the electron concentration and the electron temperature, as shown by Eq. (1.9).

16. LIGHT SOURCE AND ANALYSIS CONDITIONS

In an earlier chapter we discussed excitation processes in various light sources, as well as the effect of the discharge parameters on the concentration of excited atoms and the line intensity. A light source for spectral gas analysis can now be chosen with some understanding of the problems involved. These problems fall into two groups: those associated with the mechanism of spectrum excitation in gases and those related to changes in the composition of the mixture that take place during the discharge [365-369].

Processes associated with ***the mechanism of spectrum excitation*** will be discussed first. Most light sources will tend to excite mainly that component of the gas mixture which has the lower critical potential. This is because the temperature of the source is determined by that mixture component which has the lower ionization potential. This was shown to be true for the positive column of a glow discharge by Dorgelo [102] and Zaytsev [101]. Analogous conditions exist in an arc discharge [25]. Because of the electron temperature drop, the lines of the element with the higher excitation potential do not appear in the spectrum.

In spectral analysis of metals and alloys (and of minerals and ores) we are confronted with a similar problem of mixtures whose constituents differ in critical potentials, but the situation is not quite the same as before. For one thing, the difference between the excitation potentials is much smaller than in the case of gases.

Moreover, it is possible to use light sources in which the electron temperature is less affected by such differences. As a result, the differences in the excitation potential of the constituents do not materially affect the sensitivity of analysis, and may not even be the decisive factor. For example, the sensitivity of determination of cesium is considerably lower than that for beryllium, even though the ionization potential of cesium is much lower than that of beryllium.* On the other hand, the ionization potentials of gases differ so greatly that at low current densities many light sources, more specifically the positive column of a glow discharge, may fail to show a constituent even when the latter is present in large amounts (for example, 20-30% helium in nitrogen).

In order to improve the sensitivity of the determination of the difficult-to-excite component, analyses of these gas mixtures should be carried out in a low pressure discharge. Such light sources, however, contain nonisothermal plasmas, and thus the excitation conditions may not always be advantageous for analysis. Many experimenters have pointed out, for example, that the effect of a third component becomes very pronounced under these conditions and changes the ratio of line intensities of the minor and primary constituents. Thus the addition of nitrogen weakens helium lines [370], while the addition of argon affects both helium and neon lines [371]. Both electron temperature variations and collisions of the second kind may be responsible for such changes.

Analysis of gas mixtures involves further difficulties because the resonance lines of gases lie in the far ultraviolet, and consequently vacuum-type spectroscopic equipment is required for their recording. Today such equipment is readily available so that wavelengths below 2000 Å have been increasing in practical

---

*See [209], table on p. 57.

importance. The large amount of work on the determination of gases in metals carried out by Vodar et al. [372] was confined to that spectral region.

In the analysis of gas mixtures, the vacuum offers a double advantage over the other regions in that it makes possible the utilization of both the resonance lines of atoms and the lines produced by multivalent ions. Gas analysis in the far ultraviolet is hampered by unavailability of materials which are transparent to spectra below 1000 Å. An experimenter working with conventional spectral equipment must therefore confine himself to less sensitive spectral lines.

Light sources characterized by a high electron temperature concentration are the most suitable for the determination of the difficult-to-excite component. A high electron temperature is needed so that a large fraction of the electrons can acquire sufficient velocity to produce excitation of the difficult component (a Maxwellian electron velocity distribution is assumed). When the electron concentration is high, the excitation process is stepwise. This is evident from the following example [367]. Consider the case where the resonance line of a difficult-to-excite component ν_{01} lies in the far ultraviolet region (Fig. 62). Given a high electron concentration, the mechanism of excitation of line ν_{21}, which lies in the visible portion of the spectrum and is suitable for analysis, may be stepwise and involve a smaller energy consumption V_B' than used in direct excitation, V_B''. Stepwise excitation becomes especially significant in cases where metastable atoms are present. Stepwise excitation, furthermore, makes possible the simultaneous appearance of lines of ions of both components of the mixture. In

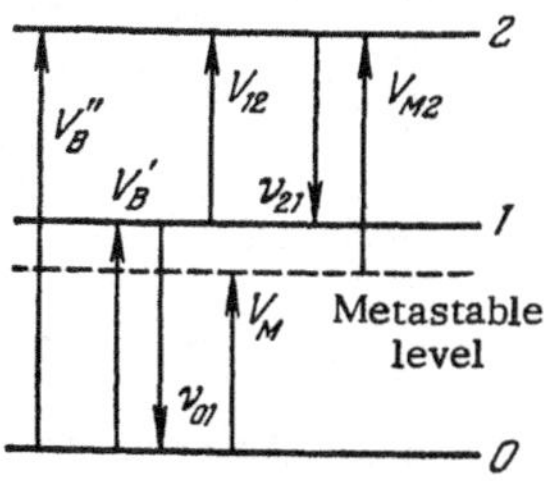

FIG. 62. Stepwise excitation of the various levels.

some instances (e.g., in the system nitrogen-argon), the ionization of both components greatly improves the sensitivity of the argon determination, because the excitation potentials of the nitrogen bands are considerably lower than those of argon, while the excitation potentials for the lines of ionized nitrogen differ very little from those for ionized argon.

We shall next consider the case of a gas mixture excited in a pulse discharge, where the electron concentration is high. By varying the discharge conditions, it is possible to excite the ion lines of the gas with the lower excitation potential, as well as the atom lines of the gas with the higher excitation potential. A requisite high electron temperature is obtained at a low pressure, using a narrow discharge tube. A high electron concentration can also be produced in the cathode fall region of a glow discharge, especially inside a hollow cathode, as well as in narrow discharge tubes and in pulsed light sources.

It follows that small concentrations of a difficult-to-excite component must be determined at low pressures, either in a high-frequency glow discharge produced in a narrow capillary, or in a hollow cathode [367]. The selection of the pressure should then be based not only on the relative intensity of the lines of the minor and principal components of the system, but also on their absolute intensities. It is therefore not advisable to use excessively low pressures since this results in an extremely faint glow. In any case, the optimum pressure must be determined experimentally. In determining small amounts of the difficult-to-excite component, one can also use pulsed sources, but the analysis should be carried out at high pressures (of the order of several millimeters). This is because in a pulse discharge a higher pressure produces a greater brightness of the flash, and hence a higher sensitivity of the determination. The method of time-resolved spectra can be used to advantage in analyzing gas mixtures in pulsed sources. This method has some clearly superior features and is now widely used in spectral analysis [273]. In analyses for small amounts of an easily excited component, a high sensitivity may be achieved by using the positive column of a high pressure glow discharge, as well as large-diameter tubes. In many cases (see Chapter V) a narrow capillary rather than a large-diameter tube is used because this permits carrying out the analysis at near-atmospheric pressures. In addition, high current densities increase the total radition brightness.

Complete qualitative analysis of a system requires a light source capable of simultaneous excitation of all the components. This is achieved in the four-electrode discharge tube of Frisch [373] (see Fig. 25b, p. 66). This tube is capable of producing three types of discharge. When electrodes *1* and *2* are interconnected and used together as the cathode, and electrode *3* is the anode, the discharge has properties resembling those obtained in a hollow cathode. With electrodes *3* and *4* one obtains an ordinary glow discharge, while the combination of electrodes *1* and *2* produces a glow discharge characteristic of closely spaced electrodes. The last type of discharge is most suitable for analyzing mixtures containing equal amounts of easy- and difficult-to-excite gases. The presence of internal electrodes is a disadvantage of this tube, since it can be used only for analysis of flowing gases.

In some cases it is convenient to install a set of various discharge tubes on the vacuum spectrograph, and to use that tube which gives the highest sensitivity in the given application.

A capillary tube closed at one end and connected to a Töpler pump (see Section 8) is also usable. In this tube the glow region can be excited over a broad pressure range (10^{-3}–6×10^{-2} mm Hg) by external electrodes coupled to a high-frequency generator.

Both the high-frequency glow discharge and the ring discharge are used in gas analysis. It was demonstrated [237] that conditions existing in a weak glow discharge are more favorable for determining the difficult-to-excite component. Considerable advantages are associated with a discharge with a frequency of the order of 10^2-10^3 Mc/s, since in this frequency range the total brightness of the discharge increases, the sensitivity of determining the difficult-to-excite component is improved [241], and gas adsorption by the walls of the discharge tube is lower [240].

Let us now consider the second group of analytical difficulties, i.e., *changes in the composition of a mixture occurring in the discharge process.* Such changes result from different rates of adsorption of the components of the mixture by the walls and the electrodes of the discharge tube. The gas adsorption is strongly dependent on the history of the tube, i.e., the duration of previous discharges; therefore, material from the same gas sample may not give the same spectrum on repeated analysis. Phenomena of this type are difficult to control. Prolonged conditioning (repeated use) of the discharge tube is the only way to eliminate the effect of previous discharges. Chemically active gases are adsorbed far more readily than inert gases, and in addition they may react with the electrode material. Such processes are all the more important since the mass of the gas is generally negligible compared with the mass of the walls and the electrodes.

To reduce gas adsorption during discharge, one must either work with a continuous flow of the gas being analyzed or use a light source operating with external electrodes. In such a source gas adsorption during the discharge is considerably reduced and the preliminary conditioning period of the tube is thus much shorter. The tube should be of a chemically stable, poorly adsorbing material. Fuzed quartz appears to offer many advantages over glass, although glass tubes have been used in many cases with good results.

A high-power pulse discharge, probably because it is short and the electrode sputtering is slight, has little if any effect on the composition of the mixture, despite the presence of internal electrodes.

Connecting an auxiliary vessel, filled with the same gas mixture, to the discharge tube helps to reduce the effect of gas adsorption on the composition of the mixture, since the adsorption loss may be compensated by diffusion of fresh portions of the mixture into the tube.

Analysis of a specific mixture may require some specific procedures. For this reason, the optimum excitation source can only be determined experimentally. The critical excitation parameters are the discharge tube diameter (which determines the current density) and the pressure. Some paradoxes do occur, however, because it has been found that the same analytical problem may occasionally be solved either by using a narrow tube at high pressures or a large-diameter tube at low pressures.

17. CALIBRATION CURVES AND SELECTION OF LINE PAIRS FOR ANALYSIS

The many methods of quantitative spectral analysis now available [25, 314, 315, 317, 374, 375] are all based on variation of intensity of a spectral line with the concentration of the element in the gas system. In most cases the concentration is determined from the relative intensities of the lines of the minor and primary constituents. Indeed, quantitative analysis was not possible until the relative-intensity method came into use. Now that stable discharge conditions can be attained and photoelectric recording of spectra is standard, new techniques designed to measure absolute intensities are being considered. In some cases analyses based on absolute intensities may indeed prove more accuate than the relative intensity procedures [361]. Nevertheless, at present it is the relative intensities that are usually measured.

According to Eq. (4.2),

$$\frac{I_B}{I_A} = \alpha C_B,$$

where I_B is the line intensity of the minor component (admixture), I_A is the line intensity of the principal component, C_B is concentration of the B, and α is a proportionality factor.

Taking the logarithm of this expression, we obtain

$$\log \frac{I_B}{I_A} = \log C_B + a, \tag{4.6}$$

where $a = \log \alpha$. If Eq. (4.6) is plotted, with the logarithm of the ratio of the line intensities of the minor and principal constituents on the ordinate and the logarithm of the concentration of the desired element along the abscissa, the graph will be a straight line with a slope equal to one.

However, the intensity of a spectral line cannot always be assumed proportional to the concentration of the element in the mixture. Thus, at high concentrations, the effect of radiation reabsorption must be taken into account (see Chapter I). In this case, the observed intensity of the spectral line is lower than the intensity of radiation emitted inside the light source because that radiation is partly reabsorbed before leaving the source. The highest reabsorption is observed in resonance lines and lines corresponding to transitions to metastable levels. Because of reabsorption, the line intensity increases with concentration at slower rates than would be expected for a linear function. We may take

$$\frac{I_B}{I_A} = \alpha' C_B^b, \tag{4.7}$$

where $b \leqslant 1$. The coefficient b characterizes the rate at which radiation is reabsorbed. When there is no reabsorption, $b = 1$. The value of b is constant only within relatively narrow ranges of concentrations (roughly, ± 500% around the mean value of the range). Reabsorption increases with higher concentration of admixture, and the value of b thus decreases. Taking the logarithm of Eq. (4.7), we get

$$\log \frac{I_B}{I_A} = b \log C_B + a, \tag{4.8}$$

where $a = \log \alpha'$. The curve of log (I_B/I_A) as a function of log C_B is called the rise curve. Over the narrow concentration range where the value of b is constant, the rise curve is a straight line with a slope equal to b [25].

It follows that within a moderately broad concentration interval, in the case where the electron temperature is not dependent on the concentration of the element in the mixture, there exists a linear relationship between the logarithm of the ratio of line intensities and the logarithm of the concentration of the element. If there is no reabsorption, the straight line expressing log (I_B/I_A) as a function of log C_B has a slope of 1, and where reabsorption is involved, the slope is b.

The relative intensities can be measured in several ways: visually, photelectrically and by photographic photometry.

Calibration curves for photographic recording

When the image density values of both spectral lines (S_B and S_A) fall within the range of the straight-line section of the characteristic of the photographic plate, it can easily be shown that

$$\Delta S = S_B - S_A = \gamma \log \frac{I_B}{I_A}, \tag{4.9}$$

where ΔS is the difference in image densities and γ is the contrast factor.

To obtain a comparison, one records on the same plate the emission of both the standard and the analyzed mixtures. The exposure conditions for all spectra being compared must be absolutely identical. Assuming that these conditions are observed, one can then obtain a calibration curve, since within the confines of a single plate γ is constant. The curve is plotted on ΔS and $\log C_B$ axes. Then one measures the difference in image densities of the lines of the desired component and the reference in an unknown mixture. The value of ΔS is computed from the calibration curve and the concentration of the desired component is determined from these data.

In principle, the calibration curve should be a straight line. However, the processes occurring during excitaton of the spectrum are so complex that departures from linearity are quite possible (see Section 15). For this reason, at least three standard mixtures must be used for plotting a calibration curve, since at least three points are needed to determine whether the curve is linear. If it is not, a greater number of standard mixtures will be needed. From Eqs. (4.8) and (4.9) we can write

$$\Delta S = \gamma b \log C_B + d, \tag{4.10}$$

where d is a constant. The slope of this curve (γb) is constant only for a particular plate, and will vary from one plate to another, depending on the type of plate and the development conditions. This procedure is known as the *method of three standards*.

An alternative procedure for plotting calibration curves is the *permanent-graph method*. The values of $\Delta S/\gamma$ are plotted along the ordinate and the logarithm of the concentration values along the abscissa. Combining Eqs. (4.8) and (4.9), we get

$$\frac{\Delta S}{\gamma}=b\ \log\ C_B+d', \tag{4.11}$$

where d' is a constant.

The calibration curve $\Delta S/\gamma = f(\log C_B)$, derived from one plate, can be used for analyzing mixtures whose spectra have been photographed on a different plate, since the ratio $\Delta S/\gamma$ does not depend on either the plate characteristics or the development conditions. In theory, a curve, once plotted, can be used for a long time; in practice, however, such curves are known to undergo parallel shifts with time [376]. It is therefore advisable, prior to using the permanent-graph method, to photograph one control standard [315], check the position of the calibration curve with it, and introduce the corrections which may become necessary should the position of the curve change (*the control standard method*).

To determine the value of γ an image density curve must be plotted for each plate or emulsion. The control standard method thus requires that each plate carry one photograph taken with a step reducer and at least two photographs of the control standard. The rest of the photographs are of spectra of the mixtures to be analyzed.

We have not mentioned here the modifications of these two basic methods for plotting calibration curves [315]. The permanent-graph procedure is advantageous when a large number of analyses is required since the labor involved in plotting the graph is then compensated by time savings achieved in subsequent analyses. However, the expenditure of this labor is not justified when only occasional analyses of a given range of compositions is required.

Graph distortions are often caused by the background effect, which 1) lowers the sensitivity of the analysis, 2) causes deviations of calibration graphs from linearity, and 3) increases the analytical error.

Assume that lines A and B appear against a fairly strong background and that the ratio of their intensities must be determined. To account for the background, its density must be evaluated separately in the spectral region surrounding the lines A and B. If the intensity distribution of background lines is nonuniform, its density should be measured on either side of the measured line and the results averaged. The intensities corresponding to the measured background densities are then found from the curve. If we set I_A and I_B as the intensities of lines A, B; I_b^A and I_b^B as the background intensities in the vicinity of lines A and B, respectively; and I_{A+b} and I_{B+b} as the total intensities of each line plus the background, then the desired intensity ratio will be given by

$$\frac{I_B}{I_A}=\frac{I_{B+b}-I_b^B}{I_{A+b}-I_b^A}. \qquad (4.12)$$

Having determined the intensity ratio for two spectral lines, we can now take the logarithm of this ratio and plot the calibration curves in the usual way, allowing for the background.

The above methods for plotting calibration curves can be used only when the plate or film is equally sensitive to and provides the same contrast for both the measured and the reference spectral lines.

In many types of work, including spectral analysis of gas mixtures, the spectral line intensities vary because not only the concentration of the elements of the mixture (including that of the desired component) changes, but also the discharge conditions vary (see Section 15). Then, from Eq. (4.5) and allowing for reabsorption, we may write

$$\frac{I_B}{I_A}=\beta' f(C_B)\, C_B^b. \qquad (4.13)$$

The exact form of the function $f(C_B)$ cannot be determined from theoretical considerations; however, within a narrow range of concentrations it may be assumed, with a fair degree of accuracy, that

$$f(C_B) = kC_B^n, \tag{4.14}$$

where k is a proportionality factor and n is positive. Hence

$$\frac{I_B}{I_A} = k'C_B^{n+b}. \tag{4.15}$$

Taking the logarithm of the above expression and rewriting it in the conventional form, we obtain

$$\log \frac{I_B}{I_A} = (n+b)\log C_B + a', \tag{4.16}$$

where a' is a constant. As we compare this equation with (4.6) we note that the slope of the straight line described by (4.16) is steeper than that of (4.6), implying a higher analytical accuracy. Indeed, it has been repeatedly demonstrated that even when a higher concentration of one of the components of the mixture alters the discharge conditions, the calibration curve generally remains a straight line. This confirms the validity of the empirical equation (4.14). Because of this linearity, three standards suffice, as a rule, for plotting the calibration curve.

Plotting calibration curves when photoelectric recording is used

In photoelectric recording the intensity of a spectral line is measured by projecting its image directly onto the photocathode of the detector photocell or photomultiplier [25, 317, 348]. The photoelectric signal is amplified in a DC or AC amplifier, so that the current from the detector is a measure of the concentration of a given element in the mixture.* In addition to being more accurate and sensitive than photographic recording, the photoelectric method saves much time and labor and is therefore indispensable for quick analysis.

In general, calibration curves based on photoelectric intensity measurements are not plotted on a logarithmic scale and consequently are not linear. The calibration curve is prepared on the basis of standard mixtures. However, several points on it must always be rechecked right before the analysis. With photoelectric recording it does not matter whether the method of three standards or the permanent-graph method is used for preparing the curve. Linear curves are obtained by replotting the data on logarithmic scales, or by restricting the calibration curves to very narrow concentration ranges when ordinary coordinates are used. The deviations from linearity of the curves caused by the presence of background radiation must be corrected, either by the methods specified in the case of photographic recording or by adjusting the detector output current to compensate for the current produced by the background. A method of photoelectric measurement where the background is automatically compensated for is described in [348].

The addition method

The addition method [209, 315, 377, 378], now widely used when working at low concentrations, must be resorted to when standards cannot be prepared because the principal substance is not available in spectral purity, or when the effect of a third-component must be compensated. There are many known modifications of this procedure. One of them, termed *the method of successive approximations*, is as follows. Let the content of the desired

---

* It is, of course, essential that the amplifier be linear.

element in the starting sample be x%. A series of standards is then prepared, with the desired element added to each in quantities C_1, C_2, ..., C_n ($C_1 < C_2 < \ldots < C_n$). The new concentrations thus will be $C_1 + x$, $C_2 + x, \ldots, C_n + x$.

In plotting a calibration curve based on the above standard mixtures the effect of the background must be eliminated. If log (I_B/I_A) is plotted on the ordinate and the logarithms of the incremental concentrations along the abscissa, the curve shown in Fig. 63

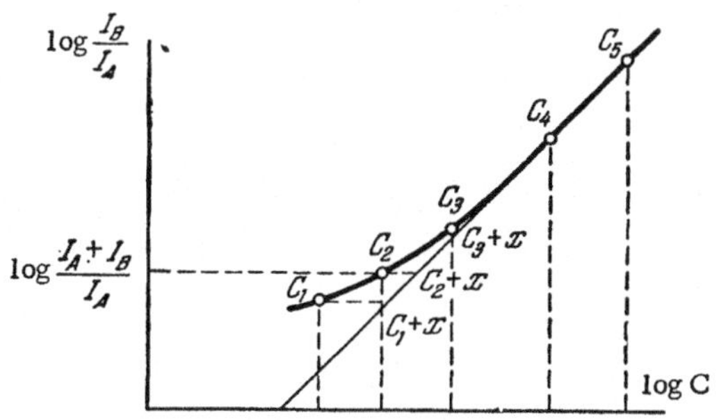

FIG. 63. Calibration curve obtained by the addition method.

is obtained. At large n the initial concentration x is negligibly small in comparison with C_n and the curve becomes a straight line (in Fig. 63, this happens at C_4). The linear section of the curve is then extended until it intersects the abscissa and is assumed to be the actual representation of the concentrations in the mixture which, together with the measured intensities, serves to determine the values of $C_1 + x$, $C_2 + x$, $C_3 + x$. Knowing C_1, C_2 and C_3, three values of x can then be calculated and averaged. This averaged value is now added to each concentration C, and the calibration curve is replotted. If the new curve is linear, it proves that x has been correctly determined. Should the curve depart from linearity in the range of low concentrations, a new correction is introduced, and the procedure is repeated until a satisfactory curve is obtained.

A recent modification of the addition method that has been adopted in ore analysis is described in [379].

The addition method should find extensive application in analysis of gas mixtures because spectral purity gases are very difficult to prepare and, since the sensitivity of analysis keeps on increasing, the requirements for gas purity will become more stringent. The addition method may thus prove the only procedure suitable for determining low concentrations of impurities or for analysis of multicomponent gas mixtures.*

The addition method is applicable to photoelectric recording if only the calibrating curves are plotted on logarithmic scales. Otherwise, the curve will deviate from linearity not only because of the presence of impurities in the starting gas mixture, but also because of the reabsorption of radiation, or because of other factors that invalidate the assumption that the line intensity is proportional to the concentration over a sufficiently wide range of C values.

Selecting analytical pairs

The accuracy and sensitivity of analysis depend to a large extent on the choice of analytical lines. If the component to be determined is present in very low concentrations (that is, in concentrations in the vicinity of the threshold of sensitivity) then obviously one has no choice and the best available analytical line is selected (sometimes only one line is available). If, however, the component is present in a concentration large enough to produce several good lines, then one should select the line that is *most sensitive* to changes in concentration, that is, the line whose

*A special modification of the addition method, useful in analysis of isotopes, was proposed by Ivanov [380].

intensity varies most markedly with changes in concentration. As was pointed out earlier, the slope of a calibration curve decreases when reabsorption of radiation is present. Therefore, the sensitivity of the analysis can be improved by using weak lines, since they are less subject to self-absorption. However, the use of weak lines is not altogether advantageous because background radiation then becomes an important factor, decreasing the slope of the calibration curve. It follows that the analytical line should be of medium intensity.

The reference line must satisfy the following requirements: 1) it must be homologous with the line of the component being sought; 2) its intensity should remain approximately constant over the range of concentrations involved and should not deviate greatly from that of the "unknown" line; and 3) its position in the spectrum must be sufficiently close to the unknown line.

Let us examine the first requirement in some detail. Lines are considered homologous if they respond to an equal extent to changes in the discharge conditions. In the case of isothermal light sources, the lines that satisfy this requirement are those emitted by atoms that are ionized to the same extent and whose excitation potentials do not differ greatly. The case of nonisothermal light sources involves one more requirement: the peaks of the excitation functions of both spectral lines must correspond to the same excitation potential.

In practice, however, analytical pairs are seldom selected on the basis of the theory. More often than not, the extent of homology is tested experimentally. The discharge conditions (current and gas pressure) are varied without changing the gas composition, and the pair of lines whose intensity ratio varies the least is selected for the purposes of analysis. Occasionally one is lucky and able to choose a pair of lines whose relative intensities do not vary even with large changes in current and pressure. Spark lines should be avoided if possible, since under conditions of stepwise excitation their intensities vary directly as the square of the current [63], and thus even slight changes in the current produce large variations in line intensity. This is not the case with arc lines, which generally reach saturation at high current densities and whose intensity thus becomes independent of the current.

If the discharge conditions vary with the concentration of desired component (for example, when determining small concentrations of an easily excited gas in a hard-to-excite gas), then the steepest slope of the calibration line is obtained with that line of the principal substance whose intensity exhibits the sharpest variations with changes of concentration of the minor component. Therefore, this line should be selected as a reference.

It is very seldom, however, that all the stated requirements can be met under actual working conditions. A compromise must be made and that pair of lines selected which ensures the highest possible analytical accuracy. The final selection is based on results obtained by testing several pairs of lines.

Another technique that has proved useful is based on introducing into the sample a foreign reference element, e.g., a metal. If the light source employs internal electrodes, one of these electrodes is usually made of this metal. In the case of gas mixtures, the auxiliary reference component is a gas, which is introduced into the discharge tube in rigorously controlled quantities (see Section 24). The lines produced by the auxiliary gas serve as reference lines. This technique is generally resorted to when the lines produced by the principal component of the mixture cannot be used as references, for example, when the principal component exhibits only a few lines in the desired spectral region, or when its ionization potential differs too greatly from that of the minor components. Differences in ionization potentials result in pronounced variations in the line intensity of the principal component with changes in concentrations of the minor ones. Thus, the lines of the principal component cannot be used as references in analysis of multicomponent mixtures.*

18. ANALYSIS OF MULTICOMPONENT MIXTURES

So far, we discussed methods pertaining to analysis of binary gas mixtures. We have seen that even in that case the relative intensities of a pair of lines are not necessarily linear functions of concentration of the minor component. The relationships being even more complex in multicomponent mixtures, the possibility of analyzing such mixtures with the excitation sources now in existence would seem to be precluded, at a first glance. This is because factors such as the strong dependence of the electric parameters of the discharge on the gas composition in the tube, the possibility of inducing uncontrollable reactions between components in the discharge, collisions of the second kind, etc., become important in that case. However, various analytical techniques have been developed specifically for work with complex multicomponent gas mixtures.

These techniques can be divided into two groups. The first is applicable in those cases where the analysis of a multicomponent mixture for each of its constituents can be reduced to the analysis of a binary mixture, i.e., where the effect of the other components on the relative intensities of a given pair of analytical lines can be virtually eliminated. This is achieved by stabilizing the excitation conditions. One of the simplest such techniques was proposed by

*In analyzing binary mixtures, on the other hand, a difference in ionization potentials between the unknown and the reference components is advantageous, since it tends to increase the slope of the calibration curve (see above).

Duffendack and Wolfe in 1939 [381] for the analysis of an H_2-N_2-CO_2-O_2 mixture. The mixture was diluted with helium (at least 1:10) and then argon was added as an internal reference. The dilution with helium greatly reduced the concentration of the various components so that the latter ceased to affect the discharge conditions. The same technique made it possible to analyze air without correcting for the effect of oxygen [382].

The excitation conditions can also be stabilized by introducing a component with an ionization potential lower than that of the desired minor constituent [383]. However, this technique, which is widely used in conventional spectral analysis, occasionally lowers the sensitivity of the analysis.

In many cases the third component does not interfer if its concentration is of the same order as that of the other minor component. This is the case in analysis for nitrogen and hydrogen in helium [384] (see Section 26).

The second group of methods consists of techniques which cope with the effect of the third and further components by means of sets of calibration curves [371] (see section 24).

The calibration curves of complex mixtures are shifted by comparison with curves for binary mixtures. The shift is caused by the component with the lowest excitation potential. Argon is such a component in an He-Ne-Ar mixture. The neon does not interfere with the determination of the argon. Therefore the argon concentration is determined by the conventional method. Then a set of calibration curves for neon intensity at various argon concentrations is prepared. This is a relatively simple case, because the third component may not only cause a parallel shift of the calibration curves, but may also change their slopes. A detailed description of specific procedures for analysis of multicomponent mixtures is given in Chapter V.

The addition method (see Section 17) may also be used: it is sufficiently accurate and sensitive to determine the concentrations of all the constituents of a complex, multicomponent gas mixture. However, this analytical procedure is extremely laborious and time-consuming; hence its use is limited to a few cases. It is employed primarily in the preparation of standard mixtures containing an impure principal component. It is further used for checking standards.

19. MICROANALYSIS OF GASES

It is sometimes necessary to determine the composition of a small quantity of a gas mixture at a very low total pressure. For example, such analyses may have to be performed when experimenting with gases liberated by metals on heating or during chemical reactions. Similar problems are involved in analyzing gases desorbed from solids, or air samples from the upper atmosphere. In such cases the analyst must work with minute quantities of the gas (of the order of hundredths of a milligram), and therefore the analytical procedure must be highly sensitive. Analysis of such minute quantities is known as microanalysis.

The methods of chemical microanalysis has been described in [385] and [386]; they require at least 1 mm^3 of the gas at atmospheric pressure. In many cases, however, the available sample consists of 1 mm^3 at 1-2 mm Hg and even less, and the techniques of chemical microanalysis are not applicable. Such minute samples may be subjected to spectral analysis, but this involves overcoming some difficulties. For example, the low pressure sample is usually in a vessel which cannot be used as a discharge tube; thus the gas must be transferred to such a tube, with all the attendant difficulties of such a procedure.

High analytical sensitivity requires that the pressure in the high-frequency discharge tube be at least 0.1 mm Hg. This pressure is achieved by either of two methods. In the first the low pressure sample is diluted with an inert gas so that the total pressure is raised to the required level. Helium is the best diluent, because it tends to increase the electron temperature in the discharge and to improve the excitation conditions for all components of the mixture. The amount of the added helium is precisely controlled and the content of all components in the helium-diluted mixture is determined. It is then possible to back-calculate the concentrations and thus arrive at the composition of the sample prior to dilution. However, addition of an inert gas introduces a new component into the sample, which complicates the analysis.

The second method of mixture analysis involves compressing the gas in the discharge tube until the required pressure is reached. The shape and size of the tube are selected to fit the specific problem. When the sample is very small, a quartz or glass

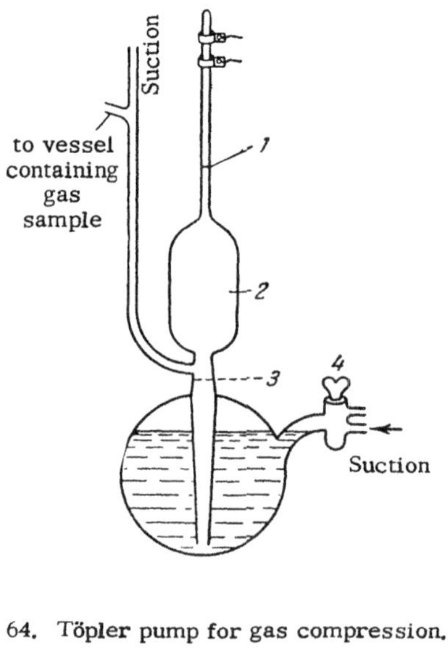

FIG. 64. Töpler pump for gas compression.

capillary closed at one end provides a suitable discharge tube [382, 387, 388]. The volume of such a capillary can be reduced to a few cubic millimeters. A Töpler pump (Fig. 64) is generally used to compress the gas. Vessel *2*, terminating in a discharge capillary, is connected to a mercury reservoir and the vessel containing the gas sample. As the mercury level is lowered to *3*, the gas fills the entire free space (vessel *2* and discharge tube, in addition to sample tube and lines). The mercury is then raised to level *1*, which pushes the gas into the discharge tube. The mercury level is raised by letting in air through a three-way stopcock *4*, and is lowered by pumping out the air. The compression ratio depends on the ratio of volumes of the discharge tube and vessel *2*.

Both methods of raising gas pressure were used by Frisch, Bochkova and Razumovskaya [382] in developing techniques for the analysis of minute air samples (see Section 25).

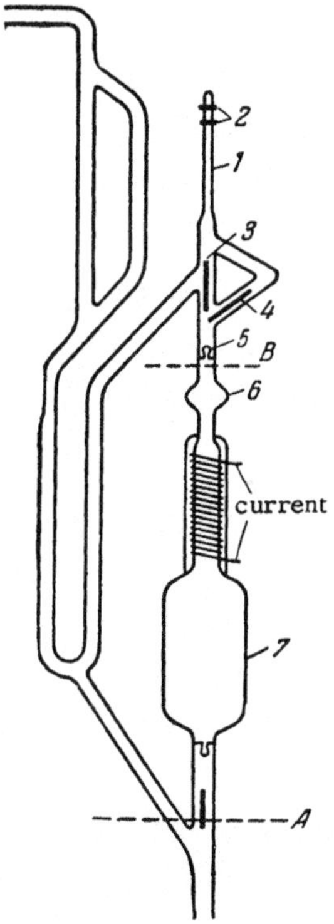

FIG. 65. Sampling bulb connected to a Töpler pump. 1) Discharge tube; 2) electrodes; 3, 4) metal hammer and removable hammer support for breaking the bulb seal at the top; 5) upper bulb seal; 6) bulb cap; 7) sample bulb; A) mercury level before breaking the bottom seal of the bulb; B) mercury level before breaking the upper seal of the bulb.

No gas will be lost if vessel *2* is eliminated and the gas sample tube is connected directly to the discharge tube. One such an arrangement (see Fig. 65) was proposed by Mirtov and is based on the break-seal valve principle* [389]. The seals on the gas

*See Brauer's Handbook of Preparative Inorganic Chemistry, Second edition, vol. I, p. 63, Academic Press, New York-London, 1963.

sampling bulb can be broken on either side. All the connecting tubing is evacuated and the mercury column in the Töpler pump is raised to level *A*. At this the lower seal of the bulb is broken. The mercury is then raised to level *B* and the upper seal of the bulb is broken. The entire gas sample is thus pushed into the capillary which serves as the discharge tube. The procedure which follows the transfer of the gas sample to the capillary (that is, the analysis proper) is virtually identical with the conventional analysis of large gas samples.

It is characteristic of spectral analysis of minute gas quantities that in most cases only one sample is available. Thus the standard mixtures and photographic conditions cannot be selected before the spectrum of the unknown mixture has been photographed. It is recommended that the first plate be used for photographing the spectrum of the analyzed mixture alone, under various excitation conditions, so that qualitative analysis can be performed. A second plate is then used to photograph the spectrum of the unknown mixture together with the spectra of the standard mixtures.

Another difficulty in microanalysis is that the pressure in the sample bulb is not always known. In this case the gas should be compressed to the limit and an attempt made to fire the discharge. If this is successful, the photographs are taken and only then is the pressure measured.* Once the pressure of the unknown mixture is known, the spectra of the standard mixtures can be photographed at the same pressure. If the pressure in the tube is too low and a discharge fails to take place, helium must be added in measured quantities until the pressure is sufficient for a discharge. The helium may be metered by means of a gasometer.

---

*It is not advisable to measure the pressure prior to analysis since this produces some loss of the sample.

When microquantities of gas are analyzed, the vacuum system must be especially clean, and greased vacuum stopcocks should be replaced with mercury valves [389].

In general, microanalysis is somewhat less sensitive than the conventional analytical procedure, since it is not always possible to select optimal conditions when only a limited number of gas samples is available. The preliminary operations involved in compression of the gas also contribute to lower sensitivity.

A reasonable accuracy in microanalysis can only be achieved if all operations are performed with the utmost care and in the correct order. Since the gas sample is very small, the gas adsorption and desorption by the walls of the discharge tube become quite significant. Because of this, the correct time interval between switching on the discharge current and the recording of spectra should be strictly observed. Photoelectric recording considerably improves the accuracy, as does the excitation of the gas mixtures with ultra-high-frequency generators.

Microanalytical procedures for air samples from the upper atmosphere have been described in [382].

Smit and Smit [42] proposed the use of Hanle's electron gun [28] for analyzing trace quantities of monatomic gases ($p = 10^{-3}-10^{-2}$ mm Hg). In an Hg-He mixture, variations of the mercury concentration do not affect the intensity of the helium lines. In such a mixture, therefore, the effect of the third component can be eliminated. For best results, single lines with broad excitation peaks should be used, since constant electron velocities are very difficult to maintain in the light source under these conditions. It would appear that only qualitative analysis is possible in this case, since a quantitative measurement would be hampered by the huge bulk of metal of the electron gun, with its attendant absorption and liberation of gases.

20. ANALYTICAL ACCURACY AND SOURCES OF ERROR

We will next discuss methods for evaluating measurement errors. A detailed description of such methods, as adapted to spectral analysis, will be found in papers [390-394] and monographs [25, 209, 315, 395].

Errors may be *systematic* and *random*. Systematic errors can be traced to some constant factor—usually defects in equipment—and will tend to distort the results of a series of measurements in a single direction: the measured values will all be either too low or too high. Random errors, as the usual implies, may cause deviations in either direction, so that the values obtained may be either high or low. Random errors may be caused by current fluctuations in the discharge tube, a defective photographic plate, and so on.

Systematic errors can be evaluated only by comparing the results of spectral analysis with those of other procedures, such as chemical analysis. However, in some types of work, such as the spectral analysis of gases, such a comparison is not always feasible. Consider one such case: let us assume that a McLeod gage, used to measure pressure during preparation of the mixture, has been incorrectly calibrated. In that case, identical results from repeated measurements will not indicate that the value obtained is accurate. They will merely show that the measurement is reproducible, i.e., that repeated analysis of the mixture yields the same data. Repeated tests provide the basis for determining the mean concentration $\bar{C}$ of a component in the mixture, which is taken to represent the true value of the desired quantity (which may possibly be distorted by systematic errors). The errors of single tests are then defined as deviations from the mean.

The criterion for evaluating a measurement method is not the magnitude of possible errors but, rather, the frequency of repetition of errors of the same magnitude. At a sufficiently large number of measurements, the distribution of the random errors will be Gaussian (or normal) [396]; that is, the number of measurements with deviation in the range $\varepsilon_i - \delta\varepsilon_i$ to $\varepsilon_i + \delta\varepsilon_i$ will be given by the equation

$$n_i = \frac{n}{\sqrt{2\pi}\,\sigma} e^{-\frac{\varepsilon_i^2}{2\sigma^2}} \delta\varepsilon_i \tag{4.17}$$

where n is the total number of measurements, σ is the root mean square or standard deviation for the given set of measurements, and ε_i is the error (deviation) of a single measurement.

From statistics, the *standard deviation* σ is

$$\sigma = \sqrt{\frac{\sum_{i=1}^{n} (\bar{C} - x_i)^2}{n-1}}, \tag{4.18}$$

where $\bar{C}$ is the (arithmetic) mean measured value and x_i is the value yielded by a single measurement (i.e., $\varepsilon_i = \bar{C} - x_i$).

The distribution of errors for two values of σ [315] is shown in Fig. 66. It is seen that at a σ of 2% measurements with errors exceeding 5% are virtually absent, while at a σ of 10% the error of a single measurement may be as high as 15% or more.

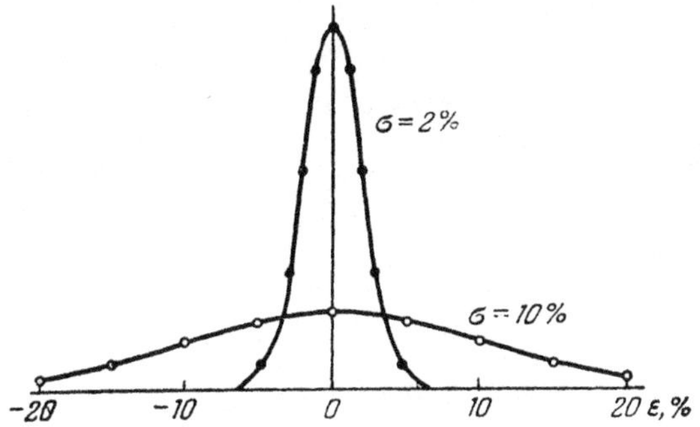

FIG. 66. Random distribution of errors.

The standard deviation may also be expressed as a percentage of the value to be determined

$$\sigma' = \frac{\sigma}{C} 100\%.^{*} \quad (4.19)$$

Alternative criteria for assessing the accuracy of a method are the *probable error*

$$r = 0.67\sigma \quad (4.20)$$

and the *arithmetic mean error*

$$\rho = \frac{\sum_{i=1}^{n} |(\bar{C} - x_i)|}{\sqrt{n(n-1)}}. \quad (4.21)$$

For a large number of measurements, the arithmetic mean error can be expressed in terms of the standard deviation as follows:

$$\rho \cong 0.8\sigma. \quad (4.22)$$

The probable error r may be simply defined as follows: one half of the total number of measurements will show deviations not exceeding r.

Any one of above definitions of error can serve as a criterion of accuracy of an analytical method.

The most frequent criterion of accuracy used so far in spectral analysis is the probable error [391]. Nalimov [393] believes the standard deviation to be more convenient, since its determination requires only a few measurements; the error distribution in this case does not have to be normal. Hereafter, we shall evaluate the accuracy of results of spectral analysis of gas mixtures in terms of this criterion.

In order to achieve higher accuracy, a measurement should be repeated several times. From theory of errors

---

*This value may be termed the coefficient of variation [393].

$$\sigma_n = \frac{\sigma}{\sqrt{n}}, \tag{4.23}$$

where n is the number of measurements from which the mean result is calculated; σ_n is the mean square error incurred when the result is taken as an average of n measurements; and σ is the standard deviation of a single measurement. We see that σ_n can be reduced by a factor of 1.7 if the analysis is performed in triplicate.

In recent years the methods of mathematical statistics have become increasingly popular. Mathematical evaluation of the results can improve the accuracy and sensitivity of analysis [395, 397]. The method most widely used today is *analysis of variance* [395], which is based on decomposition of the overall variance into a series of components, each of them accounting for a factor of the overall random error incurred in spectral analysis. Bernshteyn, Nalimov and Fal'kova [392] evaluated the precision and accuracy of spectral procedures used in analysis of geological samples by expanding the overall random error into the following terms:

1) Errors of reproducibility σ_r caused by uncontrolled factors which vary rapidly with time;

2) error σ_t due to uncontrolled factors which vary slowly with time (e.g., calibration curve shifts, variations in the slit width); it is assumed that when the tests are repeated, they are made on different days;

3) error σ_s resulting from changes in the composition of the standard mixtures.

Bernshteyn et al. determined the overall error from the equation

$$\sigma^2 = \sigma_r^2 + \sigma_t^2 + \sigma_s^2. \tag{4.24}$$

If n tests are performed on the same day, the overall root-mean-square deviation will be

$$\sigma = \sqrt{\frac{\sigma_r^2}{n} + \sigma_t^2 + \sigma_s^2}, \tag{4.25}$$

which implies that only the error in reproducibility, σ_r, is reduced by carrying out repeated tests during a single day.

In evaluating the extent to which the various component terms contributed to the total error, Bernshteyn et al. concluded that the errors in reproducibility σ_r were less significant than the other two types of error (σ_t and σ_s). This indicated that checking the accuracy of analysis by repeated photography of spectra obtained for the same mixture was of little value if the photographs were all taken on the same day. Nor can the accuracy be appreciably improved by averaging the results of several successive tests.

Let us now discuss the errors involved in the photographic methods of spectral analysis [25, 315]. Here, there are three sources of errors: 1) errors of photometry; 2) errors caused by changes in the discharge conditions; 3) errors caused by lack of uniformity in the emulsion or by differences in development. Analysis of gas mixtures involves one further error—that due to absorption (or evolution) of gases on the walls of the tube (the history of the discharge tube).

A detailed study [390] revealed that photometry errors are minor and become negligible, in comparison with the other kinds of deviation, when the test is repeated. Errors relating to the photographic plate itself are generally quite significant: even with high-grade plates and with strict observance of all instructions relating to photographic development they appear to reach 1.5-2%. When regular (as opposed to high-grade) photographic materials are used, it is not possible to reduce these errors to below 3%.

Errors associated with changes in the excitation conditions can be reduced by appropriate selection of analytical pairs (see Section 17). Irrespective of the excitation technique employed, the major discharge parameter, that is, current, must be kept constant. Voltage stabilizers in the line are thus desirable.

Analysis of gas mixtures also requires that the pressure in the sample be constant and coincide with that of the standard mixture. Pressure can be regulated by gradual removal of portions of the gas (see Section 7). The disadvantage of this method is that it is either too time-consuming (if the distance between the stopcocks is short), or does not permit precise adjustment of the pressure level. A compromise solution is usually adopted. Thus, for instance, the pressure may be controlled within 10%; however, the analytical pairs are chosen in such a manner that a ±10% pressure variation will not materially affect the intensity ratio of the line pair.

Let us now turn to the "history" of the discharge tube. In many cases the analysis is affected by the type of gas previously analyzed in a given tube. The effect is particularly marked if the light source has internal electrodes. Thus, before attempting to develop any specific analytical techniques, the experimenter must find out whether the results are influenced by the tube "history." On the basis of long experience in analyzing gas mixtures, we can recommend the following technique. When small concentrations of element B must be determined in a large amount of A, then one photographs three spectra in succession, with the content of B in A remaining the same. Each photographic experiment must be preceded by three operations:

1) Gas B is admitted to the discharge tube; the discharge is started and maintained for five minutes, after which the tube is evacuated.

2) Operation 1 is repeated. Then, gas A is introduced, the discharge is started and maintained for five minutes, and the tube is evacuated.

3) Operation 2 is repeated. Then, A is again introduced, the discharge is maintained for five minutes and the tube is evacuated. Only then are the A and B introduced.

If the ratio of line intensities of A and B is the same in all three photographs, it is concluded that the tube history has no effect on the analysis. If the above ratio is identical in the last two photographs of the set (but not in the first and second photograph), the tube should be flushed once. If the ratio varies in every photograph, the flushing must be repeated at least two more times.

A single flushing, as a rule, is not sufficient, especially since in the actual analysis the sample contains only low concentrations of B, while pure B is used in the preliminary operations. The ideal technique would involve flushing the tube with the sample, but this is not too practical, because in that case the tube would have to be flushed with the standard mixtures before taking the standard spectra. And since standard mixtures are usually prepared in the apparatus while the analysis itself is proceeding, the time required for analysis by the ideal technique would be much too long.

If the effect of previous testing in the tube is not eliminated even after a double flushing, we may conclude that the source is not suitable for analysis of a given gas mixture. It is not advisable to raise the current when the discharge tube is filled with a pure gas since this may cause hydrogen evolution from the tube walls, thereby affecting the results.

The conditioning of the discharge tube becomes much simpler if the analysis is on a flowing gas sample. This technique is now preferred to work with closed discharge tubes. On the other hand, it requires a larger quantity of sample, and the absolute sensitivity of the analysis may also be lower.

Other errors are produced in the preparation of the standard mixtures. They are primarily caused by inaccurate pressure measurements. The result is that the mixture never contains the components in concentrations coinciding with the nominal ones. It is, furthermore, difficult to prepare a mixture at a pressure adjusted precisely to the desired level. Nevertheless, just as in the case of analysis of metals, accurate estimates of the composition of standards can be obtained from the calibration curve itself [25, 315].

Let us now discuss a few suggestions for evaluating the accuracy of analysis of gas mixtures.

Obviously, we can only evaluate random errors; in other words, all we can do is to check the reproducibility of the results, that is, the precision of the analysis. Thus, several photographs of spectra of the same mixture should be taken. An alternative is to make several photoelectric measurements of the ratio of line intensities and to analyze the mixture on the basis of a calibration curve. However, there remains the matter of evaluating the effect of the previous discharges. Now, in checking the reproducibility of results obtained by a given procedure, it is desirable to keep the standard mixture at hand and admit measured quantities of it into the discharge tube. Each test with the standard mixture, whether for tube history or anything else, should be carried out with meticulous care, just as in actual sample analysis. For example, assume that it was found necessary to sweep the discharge tube with the principal component of the mixture prior to analysis. Then the tube should be flushed in the same manner before each test for the reproducibility of the results. It is reasonable to assume that the ultimate results will be improved if a large quantity of the component whose concentration in the mixture is being sought is introduced into the system *prior* to the flushing operation. After that the mixture itself is introduced, the discharge fired through it, and the tube is evacuated. Then one checks (on a sample) whether a single flushing operation is sufficient, in other words, whether or not the results differ from the other data of the series. Each new batch of the sample is subjected to several successive measurements, in order to determine whether its composition changes during the discharge. If such changes are noted, they have to be eliminated by some means, such as connecting an auxiliary vessel with the sample to the system (so that the loss of an element by absorption into tube walls can be compensated by diffusion from the auxiliary vessel), or by decreasing the discharge current. If such steps fail, we do not recommend repeated photographing of spectra of the same tube charge. In these cases, it is best to start photographing the spectra of each sample charge at exactly the same instant after the firing of the discharge, taking a single photograph. The same is true of photoelectric measurements, which in this case should also be taken at some fixed interval after the discharge has been started. As far as the calibration curve method is concerned, the reproducibility of each section of the curve should be checked separately, both in the photographic and the photoelectric method.

If a vessel with the standard mixture is not a part of the vacuum apparatus, this mixture should be prepared in the apparatus immediately before photographing the spectra of the sample. To minimize errors associated with the preparation of the standard mixture and prevent them from compounding the error in reproducibility of the results, controlled quantities of all the components of the mixture should be introduced into the system prior to each check for reproducibility, in order to keep the composition of the mixture constant. Reproducibility tests should be made one after another but over a period of several days.

There exists a simple and rapid method for evaluating the error in photographic methods of spectral analysis [25]. Instead of calculating such errors in terms of concentration, the same calculations can be carried out directly in terms of photographic density. Thus

$$\frac{\sigma}{C} = \frac{1}{0.43u} \sqrt{\frac{\sum_{i=1}^{n} \varepsilon_i^2}{n}}, \tag{4.26}$$

where u is the slope of the calibration curve; ε_i is the error of a single determination of the difference in densities:

$$\varepsilon_i = \Delta S_{\text{mean}} - \Delta S_i. \tag{4.27}$$

At present it is estimated that the results of the photographic recording method are reproducible with 3-5%, on the average. The reproducibility is highest at high concentrations in the sample and drops off as the concentrations are reduced, largely because of the interference by the continuous background spectrum. The reproducibility of analysis of gases is generally not lower than in the analysis of alloys. However, at low concentrations (tenths and hundredths of a percent) of an element, the results are reproduced (at present) to only within 7-10% of the mean.

The precision of photoelectric methods is considerably higher and, under favorable conditions, the results do not deviate more than 1%.

It must again be stressed that the checking procedures described above only evaluate the reproducibility but not the accuracy of an analysis. A reliable evaluation of accuracy must be based on comparison of results obtained on different days and in different laboratories. In addition, the results of the spectral analysis must be compared with data obtained by other analytical procedures.

21. EVALUATION AND IMPROVEMENT OF SENSITIVITY IN GAS MIXTURE ANALYSIS

In evaluating the methods of spectral analysis of gas mixtures, we shall use the concepts of absolute and relative sensitivity. *Absolute sensitivity* relates to the smallest detectable quantity of a substance; *relative sensitivity*, to its smallest detectable concentration [209].

The absolute sensitivity of gas determination is considerably higher than that of metals. Although it is lower for helium than for other gases, it is still of the order of 10^{-10} g, while in the case of nitrogen it is 10^{-12} g. The low sensitivity of metal analysis compared with that of gases is related to the low vapor pressures of metals even at high temperatures. To obtain metal vapor pressures comparable to those of gas mixtures, the samples must be thoroughly degassed and heated to high temperatures. In addition, relatively large metal samples (not less than 1 mg) are required, while a 1-mm^3 air sample at atmospheric pressure, which is more than enough for gas analysis, weighs 10^{-6} g.

However, the relative sensitivity of the determination of the difficult-to-excite components of a gas mixture is 0.1%: i.e., it is much lower than that of metals. For readily excited components, on the other hand (e.g., nitrogen in argon), the relative sensitivity is 10^{-4}%, the same as for metals.

It follows that the major problem in analyzing gas mixtures is that of improving the relative sensitivity, especially in the case of difficult-to-excite gases. Our further discussion will therefore be restricted to relative sensitivity.

The analytical sensitivity is significantly affected by the presence of impurities. In view of this, the sensitivity must be determined under the same conditions as prevail during the actual analysis. This is particularly important in qualitative analysis.

Assume we wish to test a given mixture for neon. The principal component of the mixture is helium, with nitrogen present as an impurity. Further, let us assume that preliminary experimentation has shown that neon can be detected in pure helium in concentrations of 0.001%. In the presence of nitrogen, however, the sensitivity of neon determination in helium drops off. Hence the mere fact that neon cannot be detected in a given mixture is not a proof that its concentration is lower than 0.001%. Any number of such illustrations could be cited. The problem is correctly approached by adding 0.001% neon the the sample. If a neon line then appears, the analytical sensitivity may be said to be 0.001%. Should that line fail to appear, neon must be added in greater quantities until its line becomes visible. The concentration at which neon lines appear in the spectrum of the mixture then becomes the measure of the sensitivity of the analysis.

In many cases the sensitivity is reduced by the background. The sensitivity is then closely related to accuracy. This was pointed out for the first time by Kaiser [398], who introduced the concept of the "limit of detectability" (sensitivity threshold). According to his definition, a line is detectable if the density difference between this line and the background is $3\sqrt{2}$ times greater than the standard deviation in determining the background density. It follows that the sensitivity increases with a decreasing standard deviation. If a line is to be detectable, its intensity must be several times greater than the fluctuation in the background signal. Mandelstam and Nedler [399] define the detectability threshold as that content of an element in a sample at which the signal from the spectral line of this element is 10 times stronger than the mean fluctuation of the background signal:

$$I_l\ (\text{minimum}) \sim 10\Delta I_b. \tag{4.28}$$

Nalimov, Nedler and Men'shova [397] proposed that the limit of detectability be a range, rather than a fixed concentration. The unknown concentration could then be categorized as lying either above or below this range. This approach, which the authors refer to as the "method of accumulation of measurements," gradually shifts the threshold of sensitivity into a concentration range at which a given line may record on some (but not all) spectrograms.

Regardless of the criterion for the threshold of sensitivity, the sensitivity of analysis obviously increases with the ratio $I_l/\Delta I_b$.

We shall now consider the effect of the various characteristics of the equipment on the ratio $I_l/\Delta I_b$ when photoelectric measurements are used [399, 400]. Let the dark current of the photomultiplier be low by comparison to the photocurrent produced by the continuous spectrum. Under such conditions

$$\Delta I_b \sim \sqrt{I_b}. \tag{4.29}$$

As stated earlier, in photoelectric recording one measures a light flux. The values of I_l and I_b are respectively proportional to light fluxes produced by a single spectral line and a section of the continuous spectrum. By combining Eqs. (3.17), (3.20) and (4.29), the value of $I_l/\Delta I_b$ can be derived:

$$\frac{I_l}{\Delta I_b} = k\left(\frac{D_2}{F_2}\right)^2 \left(\frac{1}{D_\varphi F_2}\right)^{\frac{1}{2}} \frac{(\Delta S_1)^{\frac{1}{2}}}{(S_2)^{\frac{1}{2}}}, \tag{4.30}$$

where k is a proportionality factor. The width of the slit S_2 can be expressed in terms of the area ΔS_1:

$$S_2 = \frac{\Delta S_2}{h} = \frac{\Delta S_1}{h} \frac{F_1}{F_2}, \tag{4.31}$$

where h is height of the slit. From (4.30) and (4.31) we obtain

$$\frac{I_l}{\Delta I_b} = k\left[\left(\frac{D_2}{F_2}\right)^2 \frac{1}{D_\varphi F_1}\right]^{\frac{1}{2}} h^{\frac{1}{2}}. \tag{4.32}$$

These expressions hold provided the exit slit does not restrict the line image. Further widening of the slit lowers the sensitivity of analysis, since it increases the background signal while the line signal remains unchanged.

In photographic recording we measure illuminance. As seen from Eqs. (3.21) and (3.27), the illuminance produced by a line spectrum does not depend on the width of the slit, while that of the stray light from the background increases with it. This may suggest the erroneous conclusion that the sensitivity increases as the slit is narrowed. Reducing the slit width in effect reduces the area of the line image, and hence the accuracy of measurement. It is therefore advisable to work with a long slit, and select the optimal width experimentally. The sensitivity of the photographic method is improved by use of high-contrast plates.

While present-day photographic methods are all based on measurements of illuminance, it would also be possible to measure the luminous flux directly (e.g., by counting the silver granules developed within the line image area).

As the exposure (that is, the time of recording of a given spectrum on photographic plate) is increased, the value of ΔI_b decreases and the sensitivity of analysis improves. It is therefore necessary to indicate the time interval required to attain maximum sensitivity. However, in the case of rapid burning of the sample larger exposure may lead to a lower sensitivity.

In both photographic and photoelectric recording the sensitivity may be increased by reducing the intensity of the background spectrum. Thus, in work with pulsed sources, one way to reduce the background intensity is to select for recording those time intervals during the pulse when the conditions are optimum for the excitation of a line rather than of a continuous spectrum.

Evaluation of the sensitivity of analysis in terms of the $I_l/\Delta I_b$ ratio allows a rational approach to the selection of the most advantageous discharge conditions. The I_l/I_b ratio must be measured as a function of the discharge parameters. The higher the value of

I_l/I_b, the larger the $I_l/\Delta I_b$ ratio; i.e., the discharge conditions become more suitable for analysis.

In many cases the analytical sensitivity is lower than that which can be handled in practice and *enrichment methods* become necessary.

Enrichment of the mixture can be accomplished by any of the physicochemical methods used in gas separation, including those used in isotope separation [401]. Such procedures, however, require complicated equipment and are time-consuming.

In some individual cases, however, special techniques can be used. For example, hydrogen easily diffuses through incandescent palladium, and oxygen through incandescent silver (see Section 10). As an example of what can be done by means of these special methods, let us discuss the case of oxygen-hydrogen mixtures. Thus, the difference between the rates of diffusion of hydrogen and oxygen through incandescent palladium can be exploited to improve the sensitivity of oxygen determination in hydrogen: in a high-frequency discharge, this sensitivity does not exceed a few tenths of a percent. By enriching the mixture with oxygen through the removal of hydrogen, the sensitivity can be improved by one order of magnitude.

To enrich such a mixture, Krasnova and Shreyder used the apparatus shown in Fig. 67. The device consists of a copper tube *1* terminating in a palladium capillary *2*. Tube *1* is connected directly to the remainder of the apparatus. The capillary is installed in a quartz tube which is wound with a wire heater supplying the required heat. The apparatus is welded to the vacuum pump connection. With stopcock *5* closed and stopcocks *6* and *3* open (and the vacuum pump on), the hydrogen is made to diffuse through the walls of the incandescent palladium capillary, and is picked up by the pump. A similar arrangement was used by Frisch and Chernyayev [402]. With the pump running continuously and a 1-mm capillary, the pressue in the system decreased from 10 to 1 mm Hg over 40 minutes. For obvious reasons, a larger capillary and higher heater current increased the rate of hydrogen removal. No oxygen diffused through the capillary. A special check was made to ascertain that no evolution of gases from the capillary walls occurred while the tube was heated. A tenfold enrichment of the mixture was thus shown to be possible.

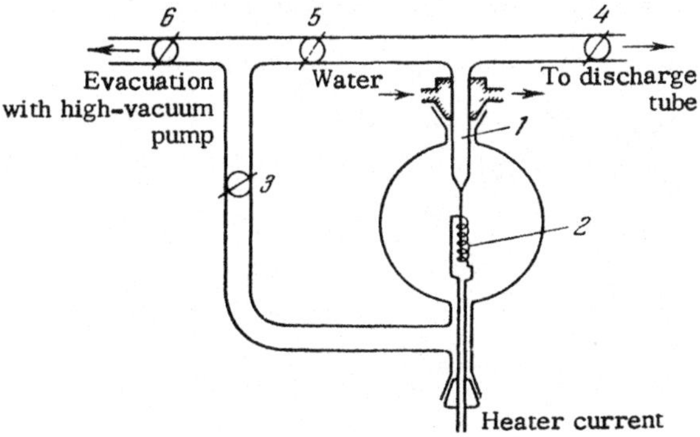

FIG. 67. Apparatus with palladium capillary for enrichment of mixture with oxygen.

An alternative mixture-enrichment technique [198, 202] is based on electrophoresis (see Section 4). As a current flows through the system, the gas with the lower ionization potential accumulates at the cathode. A better than fivefold enrichment can be achieved by selecting optimal discharge conditions and tube design. That such a procedure is worthwhile is shown by the fact that increasing the concentration of argon in an argon-helium mixture raised the relative sensitivity of the argon determination from 10^{-4}% to $4 \cdot 10^{-5}$% [202]. Unfortunately the efficiency of separation of the components by electrophoresis decreases if the mixture contains a difficult-to-excite component [203]. Hence such mixtures cannot be concentrated by this procedure. Electrophoresis is primarily used with inert gases, since molecular gases are strongly absorbed by the discharge tube walls and electrodes, so that the absorption effect cancels to some extent the enrichment of the mixture.

Enrichment can also be achieved by selective adsorption of the various gases on solid adsorbents. However, such methods must be approached cautiously, since they suffer from poor reproducibility. Nevertheless, proper selection of adsorbents allows separation of a wide variety of mixtures.* Helium and neon, for instance, can be separated from heavy inert gases by such a procedure. In some cases, the adsorption methods also allow the separation of components that lower the sensitivity of analysis, that is, removal of interfering agents.

---

*The adsorption methods are described in general manuals on gas analysis and chromatography [1, 2].

Chapter V

VARIOUS METHODS OF ANALYSIS OF GAS MIXTURES

22. QUALITATIVE AND SEMIQUANTITATIVE ANALYSIS OF GAS MIXTURES*

To start with, the light source and a discharge tube of appropriate design must be selected on the basis of considerations set forth in Section 16.

The discharge tube is sealed to a vacuum unit of the type described in Section 7. This is followed by a period of "conditioning" of the tube. This procedure is used because air is strongly adsorbed on the electrodes and the walls of the discharge tube, as are gases present in the tube before the introduction of the sample. If not removed, these gases evolve during the discharge, with the result that lines of foreign elements will appear in the sample spectrum. To avoid this, the discharge tube is filled with some gas known to be present in the sample (usually, the primary component). The discharge is then switched on for a few minutes. Its color, as well as the characteristic bands or lines in the

*See also general handbooks on spectral analysis [25, 317, 375].

spectrum, immediately indicates whether such impurities as nitrogen, oxygen, hydrogen, carbon monoxide, C_2, OH, CH or CN are present in the tube.

The current used in the conditioning of the tube is somewhat higher than is required under the analytical conditions. The gas pressure is maintained at a level approximating that required during the analysis. After this conditioning, the unit is evacuated (5-10 min) and a new portion of the conditioning gas is introduced. At first, no lines or bands indicating impurities may appear in the spectrum, but as such impurities continue to evolve from the electrodes and walls of the tube, both lines and bands will gradually be seen. The discharge is again maintained for several minutes and then the tube is reevacuated. Tube conditioning is continued until all of the lines excited by adsorbed impurities disappear completely from the spectrum. The conditioning time varies with the type of discharge. When internal electrodes are used in a glow discharge, the conditioning may require several days, but the tube, once free of impurities need not be subjected to another such long conditioning during subsequent use. On the other hand, use of a high frequency discharge tube with external electrodes requires only that the gas in the tube be replaced a few times, after which the tube is ready for use either in qualitative or in quantitative analysis.

In cases where large quantities of the sample are available, it is convenient to use the flow method of analysis. The optimum gas pressure in the discharge tube and the flow rate must be determined experimentally. The recording of the spectrum can start a few minutes after the current has been switched on, and one obtains several spectrograms in succession. The spectrogram which does not show any lines of stray impurities is used in analysis. If one uses short current bursts produced by a discharge of a capacitor, the tube need not be conditioned since neither the walls nor the electrodes heat up sufficiently to evolve adsorbed gases.

Interpretation of spectrograms consists in identifying every one of the spectral lines visible in the photograph. To speed up a qualitative analysis it is necessary to have on hand a set of photographs showing the spectra of most of the common gases, obtained on the same spectrograph that will be used in the analysis, and if possible, with the same light source. By superposing the analytical and reference spectrograms, one can identify at once a number of lines. Verifying the presence of hydrogen in a gaseous mixture is a most difficult problem, since this gas evolves not only from the walls of the tube but also from the greases and sealing compounds. This is especially true during the first few days after regreasing of stopcocks when the persistent hydrogen line, H_α, can be detected in virtually all spectrograms. Rigorous precautionary measures are therefore necessary when determining traces of hydrogen; specifically, it is advisable to work with weak currents. The spectrum of the gas sample is photographed along with the spectrum of its primary component. The presence of hydrogen cannot be unequivocally established unless the analytical spectrogram displays greater intensity of the hydrogen lines than the reference photograph.

Historically, the first paper on qualitative analysis of gas mixtures, which dates back to 1896, described the determination of hydrogen in air [403]. Then Moureu and Lepape [404] determined the krypton and xenon concentrations in argon by visual photometry. They adjusted the pressures so as to obtain a match between the intensities of the krypton and xenon lines and those of the argon lines at various concentrations of krypton and xenon in argon. Tables of homologous line pairs were then compiled for the concentration range of 0.05 to 0.5%. Pentcheff [405] used the

Moureu-Lepape technique for the analysis of gas from water originating from various natural sources and improved the sensitivity of the method, achieving a limit of detectability of 0.01%.

Heyes [406] worked with condenser spark discharges between metal electrodes at atmospheric pressure. The gas mixture was introduced by suction into a discharge tube in which aluminum electrodes were mounted 8 mm apart. An induction coil (this was the excitation source in many early experiments) with the breakdown gap of 15 cm supplied the current. A capacitance $C = 8260\ \mu F$ and an inductance $L = 2.3 \cdot 10^{-5}$ henry were connected in the circuit of the secondary, with a 7 A current flowing through the primary. Analysis, carried out by the method of persistent lines as well as by the method of homologous pairs (internal standard), yielded the following results:

Sample	Concentration range, %	Analytical method
Nitrogen in hydrogen	0.037-7.8	Persistent lines and homologous pairs
Hydrogen in nitrogen	1-20	Homologous pairs
Nitrogen in oxygen	0.057-12.5	Both methods
Oxygen in nitrogen	1.25-12.5	Both methods

The intensities of nitrogen lines were compared with those of iron lines, since the electrodes contained iron as an impurity. However, this procedure is not too good, because the intensity of lines of the electrode material varies (with the discharge conditions). As seen from the table, the sensitivity was poorest in the case of determinations of impurities in nitrogen. The sensitivity of hydrogen determinations in nitrogen is improved if helium is added. This sharply increases the intensity of the H_α lines. Thus, the intensity of the H_α line at a concentration of 1.75% H_2 in N_2 is the same as when 10% H_2 is present in the absence of He. The Heyes method is still extensively used in spectral analysis of gas mixtures, because addition of large quantities of an element with a high excitation potential increases the electron temperature in the discharge gap and, consequently, the analytical sensitivity for the difficult-to-excite components.*

Somewhat earlier, interesting results were reported by Paneth and Petters [407] during analysis of helium and neon in air. They used an induction coil connected in parallel to a spark gap. The discharge tube was a narrow capillary. The thin wire electrodes could be moved in the capillary, thus materially altering the discharge conditions.

In later experiments, Paneth and Günther [408] abandoned the induction coil, because the method failed to produce spectral lines of sufficient intensity. They then worked with high frequency electrodeless discharges from a Tesla transformer and determined small quantities of hydrogen and neon in helium. The gases were excited in 0.1-0.4 mm capillaries. Hydrogen was detected in helium in concentrations as low as 0.1%, with the appearance of H lines being strongly dependent on pressure. Best results were obtained at pressures ranging from 2-4 mm Hg. In determining neon in helium, the pressure range of 2-4 mm Hg was again used, and the limit of detection was 0.002%.

Qualitative analysis of inert gases in helium is described by Karlik [358]. An h-f tube oscillator was used for spectrum excitation [409]. The quartz discharge tube, 1-1.5 mm in diameter, carried external electrodes set 3.5 cm apart. The range of pressures investigated was 0.01 to 0.1 mm Hg. The tube was connected to the remainder of the unit by means of a mercury seal so that it could be rotated alternately toward one of two spectrographs, one

---

*The optimum quantities of helium added to the mixture were selected by experiment. It would appear that it is undesirable to dilute the mixture more than by a factor of 10.

of which was used for the visible and the other the UV spectrum. Long low pressure discharge was associated with a fatigue effect, which was apparent from the fact that it became more and more difficult to start the discharge. The fatigue was removed upon introduction of air or oxygen into the tube. Prior to analysis the tube was conditioned in pure helium. Karlik tabulated the limits of detection (in %) for argon, krypton, xenon and neon in helium, both in the visible and the UV regions:

Gas	Spectral region		Absolute quantity cm^3
	Ultraviolet	Visible	
Argon	$5 \cdot 10^{-2}$	$3 \cdot 10^{-2}$	$6 \cdot 10^{-9}$
Krypton	$3 \cdot 10^{-3}$	$2 \cdot 10^{-3}$	$4 \cdot 10^{-10}$
Xenon	$3 \cdot 10^{-3}$	$3 \cdot 10^{-3}$	$6 \cdot 10^{-10}$
Neon	$2–6 \cdot 10^{-3}$	$2–3 \cdot 10^{-3}$	$5 \cdot 10^{-10}$

Traces of a difficult-to-excite impurity of a gas mixture can be detected in the cathode regions of a glow discharge. Thus, Frisch and Konovalov [366] were able to detect a few tenths of one percent of argon in nitrogen.

To determine argon in nitrogen, Van Liempt and Visser [410] photographed radiation emitted by the positive column of a glow discharge and studied the argon line Ar λ 4159 Å. When the glow is photographed in the vicinity of an internal electrode, the sensitivity for the difficult-to-excite component in a high-frequency discharge is no higher than in analysis confined to the positive column of a glow discharge. However, with external electrodes the sensitivity can be improved to a point where argon can be detected in nitrogen at concentrations of a few tenths of one percent.

Monfils and Posen [411] developed a method for determining traces of nitrogen in argon. The analysis was carried out in a hollow cathode and in the anode portion of the discharge, using a pointed anode. In using the hollow cathode portion and pressures of the order of 10^{-1} mm Hg, the authors were able to detect argon in concentrations of the order of 10^{-2}%. The authors noted that, while the sensitivity for nitrogen increases with pressure in the tube, the discharge becomes unstable. In the anode region, the limit of detection was 0.01%. However, neigher in the anode region nor in the hollow cathode were the discharge conditions favorable for excitation of nitrogen in argon.

Duffendack and Wolfe [381] obtained high sensitivity for nitrogen and hydrogen in helium. In the positive column of a glow discharge, hydrogen was detectable in concentrations as low as 0.005%, and nitrogen in concentrations of about 0.02%. (In a high frequency discharge, the limit of detection was of the order of 10^{-4} %.)

In a pulse discharge [271], helium can be determined in nitrogen or argon, and neon in air, when present in concentrations of the order of 10^{-2}%.

The above summary shows that, given a judicious selection of the excitation source, discharge conditions and the discharge tube dimensions, the limit of detection for the difficult-to-excite component can reach 10^{-1}%, and in some cases even 10^{-2}%, whereas the same limit for the easily excited component of the mixture is of the order of 10^{-4} %.

Spectrograms for qualitative analysis can be obtained readily and speedily with the aid of a very compact device described in [412]. This device consists of a pocket-size Zeiss direct vision spectroscope combined with a camera for recording the spectrum on film (see Fig. 68). The excitation of the gas mixture is achieved by means of a standard 2450 Mc, 10-100 W magnetron oscillator.

In many cases, it is relatively simple to obtain a rough quantitative estimate of the composition of a mixture. This is done

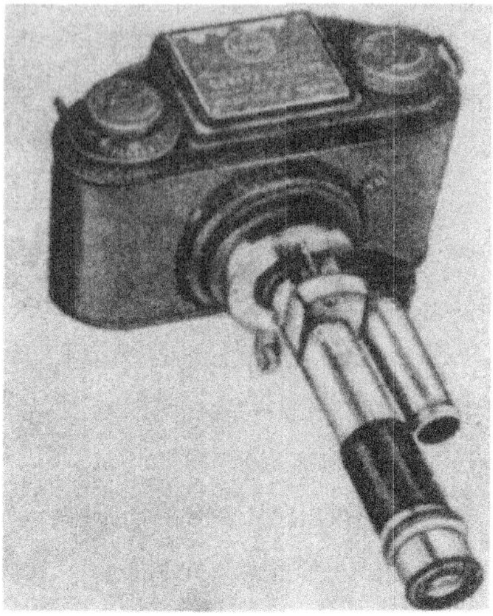

FIG. 68. General view of a spectroscope camera for qualitative analysis.

semiquantitatively either on the basis of persistent lines or with homologous pairs. The accuracy of the method of persistent lines is not better than the order of magnitude, since the discharge conditions must be very exactly reproduced. More accurate results are obtained via the method of homologous pairs. As a preliminary step, one prepares a table of lines of equal intensities at various concentrations in the mixture. This is done at strictly specified discharge conditions which are then maintained constant. This method can be used both with photographic recording or visual examination of the spectrum. At the present time, it is used mainly in semiquantitative analysis performed with the aid of a styloscope.

When checking the purity of a gas, it is often sufficient to establish only the upper or lower cncentration limits for such impurities present. This amounts to, in effect, to a visual semiquantitative evaluation of the impurity content, an evaluation which can be successfully accomplished by means of a styloscope or some other type of visual spectroscope. The visual methods of semiquantitative metal and alloy determinations by means of a styloscope or stylometer are well developed [413] and are widely used in the industry. Styloscopic determinations of gas mixtures are, in fact, less complicated than those of alloys because gas spectra show fewer lines. Thus, in an actual case of gas mixture analysis with the aid of a styloscope, minute quantities of neon were determined in helium [414]. A high-frequency oscillator was used to generate the discharge. The discharge tube (8 mm I.D.) was made of quartz and contained the sample at a pressure of 8 mm Hg. The light emerging from the discharge capillary was allowed to fall directly upon the styloscope slit, obviating a condenser lens. The most sensitive neon line in the visible spectrum was the λ 6402 A line. The only helium line displayed in this region (λ 6678 A) was used as the comparison line. When neon is added to pure helium, its line λ 6402 A begins to appear in the spectrum only at concentrations of 0.001% or greater. When neon is present in a concentration of 0.08%, the neon and helium lines show equal intensity. Helium spectra are characterized by the scarcity of lines, hence it is extremely difficult to find helium and neon lines of equal intensity at low neon concentrations in the sample. For this reason, the relative line intensities of neon and helium were compared by viewing the spectrum through an optical step attenuator, set up in front of the styloscope slit and at some distance from it. If the attenuator is placed directly on the slit, then the astigmatism of the instrument causes the vertical spectral lines and the horizontal steps of the filter to focus in different planes, so that it is impossible to see clearly and simultaneously both the spectrum and the individual steps of the attenuator. By placing the attenuator in front of the styloscope slit, it is possible to focus in one plane the individual steps of the attenuator and the spectral lines. Comparison of neon and helium line intensities gave the following results for different degrees of attenuation:

Neon concentrations, %	Steps of the optical attenuator (on the left, Ne λ 6402 Å; on the right, He λ 6678 Å)
0.001	The 6402 Å line becomes visible
0.009	1 = 2
0.014	1 = 3
0.030	1 = 4
0.050	1 = 5-6
0.070	1 = 7
0.080	1 = 1
0.095	8 = 1
0.150	7 = 1
0.220	5 = 1
0.470	3 = 1
0.940	2 = 1
Logarithms of the transmission values for the attenuator steps are: 2.0; 1.86; 1.74; 1.64; 1.51; 1.38; 1.24; 1.12.	

The concentration range of 0.001–0.01% can be calibrated if a step attenuator of greater optical density is available. Styloscopic determination of neon in helium thus becomes possible at concentrations ranging from 0.001 to 1%. However, other analytical line pairs must be used to determine higher neon concentrations, since at those concentrations the neon line λ 6678.28 Å superposes on the helium line λ 6678.15 Å. When neon is present in low concentrations, this neon line is weak and cannot affect the results. At neon concentrations of 1% and higher, the analysis should be based on the line pair He λ 5876 Å – Ne λ 5852 Å.

The determination of the composition of other gas mixtures by the visual semiquantitative technique is also relatively simple.

23. QUANTITATIVE ANALYSIS OF BINARY GAS MIXTURES

The analysis of binary gas mixtures can be divided into the following groups:

I. Determination of molecular gases in inert gases.

II. Analysis of mixtures of inert gases.

III. Determination of inert gases in molecular gases.

IV. Analysis of mixtures consisting of two molecular gases.

The first group of analysis can be classified as determination of an easily excited component in a difficult-to-excite primary constituent, while the third group, is the opposite, that is, determination of a difficult-to-excite minor component in an easily excited primary constitutent. In groups II and IV, the primary constituent can be either an easily or a difficult-to-excite gas.

We shall discuss a few typical cases which illustrate the analytical approach to each of the above groups. The gas mixtures as well as the analytical pairs used are listed in Appendix II.

I. Determining nitrogen, oxygen, hydrogen and hydrocarbon impurities in inert gases

This is the simplest case of analysis of an easily excited component in a difficult-to-excite gas, and there are no major difficulties in achieving relatively high sensitivity. The light source is the positive column of a DC or a high-frequency discharge. The most reproducible results are achieved with a high frequency discharge in a tube with external electrodes. In the determinations of nitrogen, oxygen and hydrogen in inert gases, the maximum relative sensitivity (of the order of 10^{-3}-10^{-5}%) can be obtained only when the gas pressure in the discharge tube is high. Increasing the pressure in large-diameter discharge tubes causes a compression of the positive column, which results in a significant decrease in the brightness of the radiation. Therefore, there is nothing to be gained by operating at pressures higher than 20 mm Hg, because the relative sensitivity under these conditions does not exceed a few hundredths of one percent. On the other hand, in narrow capillaries (0.5-1 mm bore), pressures of several hundred mm Hg do not reduce the brightness of the emitted radiation, so that one can determine impurities in concentrations of 10^{-3}-10^{-5}% [415].

Nitrogen determination in helium

Nitrogen determinations in inert gases require special attention to the condition of surfaces in the discharge tube. Nitrogen adsorption from the sample has been observed even in discharge tubes subjected to efficient preliminary conditioning. It was noted, for example, that nitrogen bands in helium spectra become detectable only at concentrations of about 10^{-3}%. However, if the tube has been flushed several times with a mixture containing 10^{-3}% nitrogen in helium, the intensity of nitrogen bands will increase appreciably upon addition of only $2 \cdot 10^{-5}$% N_2 to this mixture. The calibration curves for nitrogen in helium (see Fig. 69) were obtained under

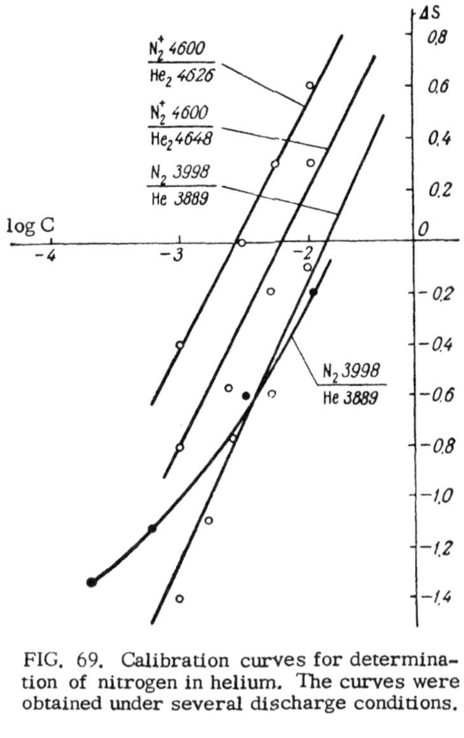

FIG. 69. Calibration curves for determination of nitrogen in helium. The curves were obtained under several discharge conditions.

the following conditions. The mixture was excited by a high frequency discharge (f = 6 Mc) in a 1 mm I.D. capillary, at a pressure of 600 mm Hg and a current of 300 mA. Under these conditions the visible spectrum of the mixture (see spectrum X in Appendix VI) displays lines of atomic helium as well as bands of molecules He_2, N_2 and N_2^+. Analytical pairs such as He_2 λ 4648 Å—N_2^+ 4600 Å, He_2 λ 4626 Å—N_2^+ λ 4278 Å, N_2 λ 3998 Å—HeI 3889 Å may then be used. Pressure variations of 100 mm Hg do not affect the variation in the relative intensities of the He_2 and N_2^+ bands. Analytical results are reproducible to within 5%.

Nitrogen determination in neon

Over the concentration range of 10^{-4} - 10^{-2}%, nitrogen admixtures in neon are determined from the bands of the second positive nitrogen system $C^3\Pi - B^3\Pi$ lying in the 3500 Å region (see Appendix VI, spectrum XI). At nitrogen concentrations in the sample of the order of one tenth of one percent, the first positive system of nitrogen bands $B^3\Pi - A^3\Sigma$ can likewise be used for nitrogen determination.

If the nitrogen-containing sample also containing small quantities of carbon dioxide and hydrocarbons, the nitrogen can be determined in neon from the CN bands, whose intensity varies linearly with changes in nitrogen concentration. The analytic conditions are the same as in the case of nitrogen in helium or argon. The mixture is excited at a pressure of the order of 100 mm Hg, using a high frequency discharge and a 0.5-1 mm I.D. capillary. An appropriate interference filter can be used to separate the neon bands.

Nitrogen determination in argon

The analytical procedure for determination of nitrogen in argon with a high frequency discharge as an excitation source is described in Section 26.

If the sample is flowing through the discharge tube, the nitrogen-argon mixture can be excited not only in a high frequency discharge, but also in the positive column of an AC discharge [416] or in a hollow cathode [366]. However, the order of magnitude of the relative sensitivity of nitrogen determination in these sources does not exceed 0.01%, with the average analytical error amounting to about 20%.

When traces of nitrogen are determined in a hollow cathode it is advisable to base the analysis on the λ 4236 Å band. At nitrogen contents ranging upward from 0.1%, the λ 4709 Å band can be conveniently employed. The use of analytical pairs Ar λ 3949 Å—N_2 λ 3943 Å or Ar λ 4044.42 Å — N_2 λ 4059.4 Å is recommended by one experimenter [416].

It was shown [417-420] that the determinations of traces of nitrogen in argon can be conveniently based on the total intensity of bands II of the positive nitrogen system. These bands are located in the 3900-3600 Å region, which is easily separated by means of filters. This technique has been employed by a number of authors in developing automatic gas analyzers [358, 359, 421, 422]. A description of such analyzers was given in Section 14.

Oxygen determination in inert gases

Oxygen concentrations of 10^{-4} % and higher can be determined in neon and helium if the sample is excited by means of a high frequency discharge in 0.5-1 mm I.D. capillary at a pressure of $\sim$ 600 mm Hg. The analysis is based on the lines of atomic oxygen in the infrared spectrum. When neon or helium are the principal constituents, the analytical line pairs are OI 7772 Å—NeI 7839 Å and OI 7772 Å—HeI 7281 Å, respectively.

Again, the discharge tube should be flushed with the sample prior to analysis, since oxygen is adsorbed on the discharge tube to an even greater extent than nitrogen.

The limit of detection of oxygen in argon is lower than in helium or neon, it being no higher than 0.01%. The analysis should be based on the analytical pair OI λ 7772 Å—Ar λ 7624 Å.

Figure 70 shows the calibration curves for oxygen in inert gases. The shallow slope of the curves and their deviations from linearity over the range of low concentrations can be attributed to the presence of a background as well as traces of oxygen which remain adsorbed on the walls of the tube even after flushing with pure inert gas.

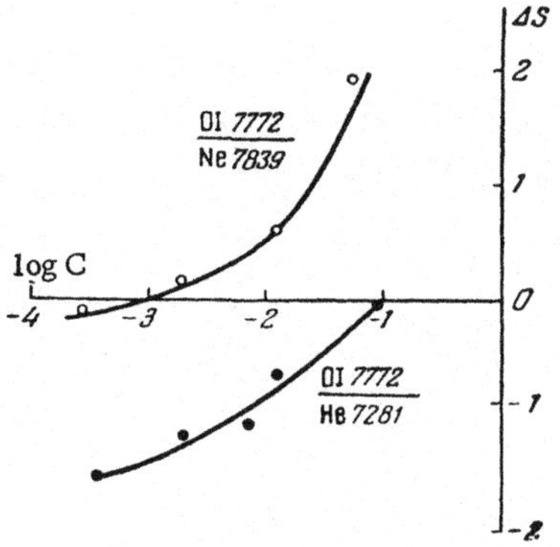

FIG. 70. Calibration curves for determination of oxygen in neon and helium.

The sensitivity of oxygen determinations can be improved by using equipment of high dispersion. The above results were obtained with an ISP-51 spectrograph, f = 270 mm (see Appendix IV). Simplified analytical techniques can be used provided one also employs a monochromatic filter with a very narrow passband for the OI λ 7772 Å line.

Hydrogen determination in inert gases

In this case, both the vacuum unit and discharge tube should be thoroughly outgassed, and all water vapor should be removed. Outgassing the equipment does not remove all the water: over 18-25 hours, the amount of water vapor evolving from the walls of even a carefully outgassed tube is sufficient to increase the hydrogen content of the sample by several hundredths of one percent (the discharge dissociates the water). In addition, the release of water vapor from the walls is accelerated by the discharge, and the rate of evolution increases with the discharge current. To add to the difficulties, the tube adsorbs hydrogen from the sample. Thus, if the sample in the tube is replaced by a pure gas, the latter will show traces of hydrogen. The latter cannot be completely removed unless the tube is heated by means of a gas burner. These complications are encountered with all mixtures containing hydrogen.

Hydrogen analysis in helium is described in [384, 419, 423, 424]. A helium-hydrogen mixture is excited by a high frequency discharge in a 1-2 mm I.D. capillary, the maximum working pressure being ~20 mm Hg. The limit of detection can be as low as 10^{-3}% [384]. Hydrogen determinations in helium can be carried out both in sealed and open discharge tubes, under conditions of continuous passage of the gas [384, 423]. The analtyical line pairs used are $H_\beta \lambda$ 4861 Å—He λ 4922 Å, $H_\beta \lambda$ 4861 Å—He λ 5047 Å. The red hydrogen line $H_\alpha \lambda$ 6562 Å can also be used. In the concentration range of 10^{-3}-10^{-2}%, the analytical errors amounted to about 25% and to 10% for concentrations ranging from 10^{-2} to 10^{-1}%.

To obtain reproducible results, especially at low hydrogen concentrations, the discharge tube should be thoroughly conditioned by heating with a gas burner after each analysis [384]. When working in a stream of gas, the conditioning of the tube is done with the sample during preliminary firing.

According to [419], an alternative method of hydrogen determination in helium consists of isolating the analytical lines by means of monochromatic interference filters. The limit of detection is in this case determined by the quality of the filter used to separate the hydrogen line. Filters with λ_m = 486 millimicrons, $\Delta\lambda$ = 10 millimicrons at T_r/T_m of the order of several percent, allow the determination of hydrogen in helium when the former is present in amounts of not less than a few hundredths of one percent.

The methods of hydrogen determination in argon and neon are the same as those used in analyzing helium for its hydrogen content.

Determination of total hydrocarbon content in inert gases

Emission spectroscopy can only determine the total hydrocarbons in an inert gas, without identifying the compounds. Analysis is based on the CH λ 4315 Å or C_2 5165 Å bands, which are easily excited in any type of discharge.* The limit of detection ranges from 10^{-4} to 10^{-5}%. However, if the hydrocarbon content is greater than 0.01%, the analysis becomes virtually impossible, because the spectrum of the sample shows nothing but C_2 and CH bands whose intensities do not vary with concentration of the hydrocarbon. If the inert gas also includes hydrogen, the hydrocarbon analysis becomes ambiguous, since the intensities of C_2 and CH bands become greater as the concentration of the hydrogen impurities in the sample increases. In such cases, one can establish only the range of the hydrocarbon content in the sample, and even that, more often than not, will involve large errors.

In these paragraphs we discussed only the determination of single minor components of samples consisting mainly of an inert gas. Such analyses are seldom encountered in practice, because what is generally required is the simultaneous determination of several components.

Various techniques for simultaneous determination of several components of a sample are described in Sections 18, 25 and 26.

II. Analysis of mixtures of inert gases

Analyses involving binary mixtures of inert gases are required in the manufacture of these gases. For instance, helium of spectral

*The vapors of the sealing compound (grease) interfere in this analysis because they are decomposed in the discharge, producing high-intensity CH bands.

purity is known to invariably contain traces of neon which cannot be removed by the usual methods. Conversely, neon always contains some helium. It is also very difficult to separate krypton and xenon. Furthermore, a krypton-xenon mixture invariably contains traces of argon.

Determination of neon in helium

Neon can be quantitatively determined in a helium-neon mixture when present in concentrations ranging from 0.001 to 99.9%. The standard deviation of a single analysis varies between 3.5 and 7.5%, depending on the concentration. Figures 71 a and b show calibration

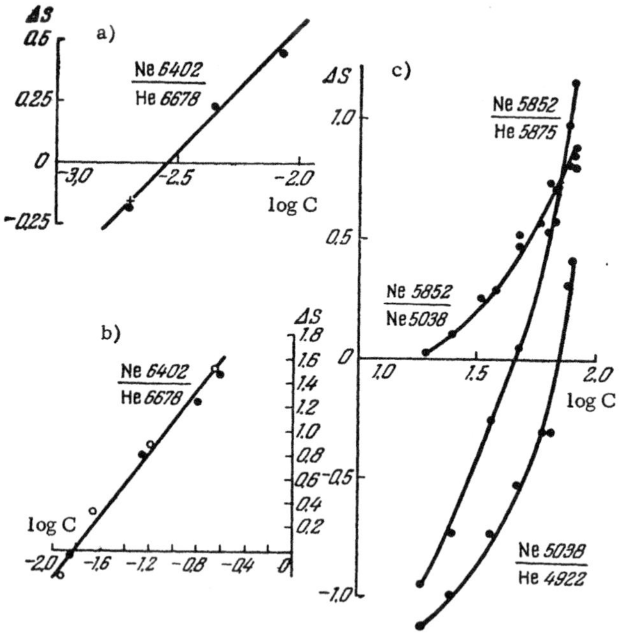

FIG. 71. Calibration curves for determining the neon concentration in a neon-helium mixture.

curves for determination of neon in helium when the former is present in low concentrations (0.002 to 0.01%). The spectra were obtained in a high-frequency discharge using an 8 mm I.D. capillary, a pressure of 10 mm Hg, and a discharge current of 200 mA. The analytical line pair used was Ne λ 6402 Å—He λ 6678 Å. Figure 71c shows calibration curves plotted on the basis of various analytical pairs for neon concentrations in helium ranging from 18 to 75%. It is evident from this graph that electron temperature variations associated with changes in the composition of the mixture alter the ratios of intensities of different lines of one of the components. Thus, when neon is present in helium at a concentration of 18%, lines Ne λ 5852 Å and Ne λ 5038 Å have about the same intensity, while at a concentration of 75%, the first line is far more intense. Because of this, there is no need for using the lines of a second element as a reference, and the analysis of a mixture can be based on calibration curves plotted from the relative intensities of different lines of the same element, which are changed to a different extent by variations in the discharge conditions. The most conspicuous variation of intensity ratios is noted in the arc and spark lines of one and the same element.

The relative sensitivity of quantitative analysis of helium-neon mixtures is lower than that for helium in nitrogen. This is because an increase in the total pressure in the discharge tube to 600 mm Hg drastically alters the spectral pattern of helium. Thus, at pressures of 100 mm Hg and higher, the spectrum shows, in addition to atomic lines, high intensity He_2 bands (6399, 6310, 5862, 4649, 4626 Å) slightly shaded to the red region of the spectrum. The He_2 λ 6399 Å band becomes superimposed on the sensitive line Ne λ 6402 Å. For this reason, the pressure in the discharge tube should not exceed 100 mm Hg for neon determinations in helium. The limit of detection for neon in this case is 10^{-3}%.

The procedure of neon determination in helium is readily simplified if the neon lines are separated by means of monochromatic filters with a broad passband for 200 Å and λ_m 6400 Å [419].

The limit of detection of helium in neon is 0.1% or less. The 5876 Å analytical line for helium is very close to the neon line 5882 Å. For this reason, the simplified analytical procedure based on the use of filters is not very sensitive. For direct determination of neon in helium in gas discharges, see Section 27.

Xenon determination in krypton

This analysis involves no particular difficulties. Limits of detection of about 10^{-2}-10^{-3}% can be attained, if the sample is excited in the positive column of a high frequency discharge (1-5 mm I.D. capillaries and pressures of 5-10 mm Hg).

The determination of traces of krypton in xenon is considerably more complicated. The limit of detection, even with a high frequency discharge produced in 0.5 mm I.D. capillary at pressures of 0.4-0.5 mm Hg, does not exceed a few tenths of one percent. The analysis is based on the analytical line pairs Kr 5871 Å—Xe 5875 Å, Kr 5871 Å—Xe 5976 Å, Kr 5571 Å—Xe 5419 Å, etc. Calibration curves for krypton determination in xenon are shown in Fig. 72.

Krypton detemination in xenon can be based on the electrode regions of the discharge, or the glow inside the external electrode (which is an analog of a hollow cathode in a high frequency discharge). Both arc and spark krypton and xenon lines in the 4000-4800 Å region can be used as the analytical pairs (see Appendix V, spectrum V).

Argon determination in a krypton-xenon mixture

Determination of traces of argon in a krypton-xenon mixture, as well as of traces of krypton in xenon, are typical examples of analysis for a difficult-to-excite component in an easily excited primary constituent. There is little difference between the excitation potentials of Kr and Xe lines, but argon lines require considerably higher excitation potentials. A further difficulty associated with the

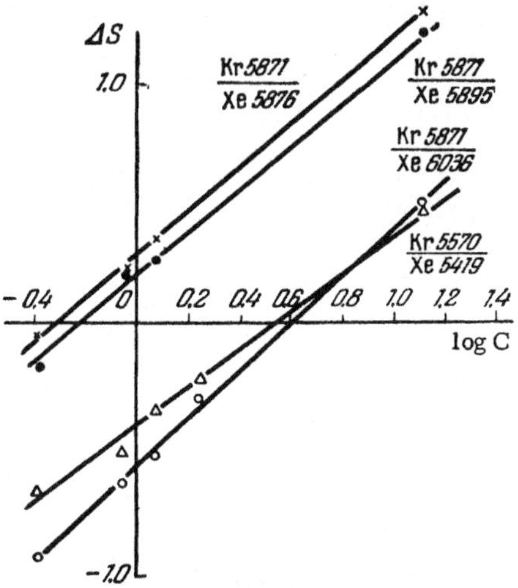

FIG. 72. Calibration curves for krypton determination in xenon.

determination of argon in krypton-xenon mixtures lies in the fact that the sensitive argon lines which appear during transitions to metastable levels $^3P_2$ and $^3P_0$ are located in the infrared region. Nevertheless, a high frequency discharge, using a 0.5 mm I.D. capillary and pressures of 0.3-0.5 mm Hg, allows determination of argon in concentrations of a few tenths of one percent. The analytical line pairs used are Ar 6965 Å—Kr 6905 Å, Ar 7067—Kr 6905 Å. In the concentration range of 0.3-0.1%, the calibration curves have slopes close to unity, while for concentrations ranging between 1-20%, the slopes are greater than 2.* The error of a single determination does not exceed 10%.

III. Determination of inert gases in molecular gases. Determination of argon in nitrogen

Frisch and Konovalov [366] were the first to attempt the analysis of a nitrogen-argon mixture. They used a discharge tube with a hollow cathode (see Fig. 25b).

---

*The large slope is apparently due to the fact that the intensity of lines of the primary constituent also varies with the concentration of the minor component.

As we pointed out earlier (Section 4), a tube with a hollow cathode tends to confine the entire glow region to the interior of the cathode, where both the bands of the molecular nitrogen and the spark lines of argon are excited simultaneously. Frish and Konovalov compared lines ArIIλ 4765 Å and ArIIλ 4736 Å with a line in the band λ 4709 Å in the rotational structure of nitrogen. The ratio of intensity of the nitrogen band to these argon lines was shown in special experiments to be practically independent of both the current and pressure. Figure 73 shows the calibration curve for argon concentrations in nitrogen ranging from 2 to 96%. The spectra were recorded at 0.65 mm Hg and 65 mA. The analytical sensitivity is very high at low concentrations, where the curve is steep, but it is not so high in the middle region where the slope is shallow. When using a hollow-cathode tube, one can detect argon in nitrogen at concentrations of the order of 0.1%.

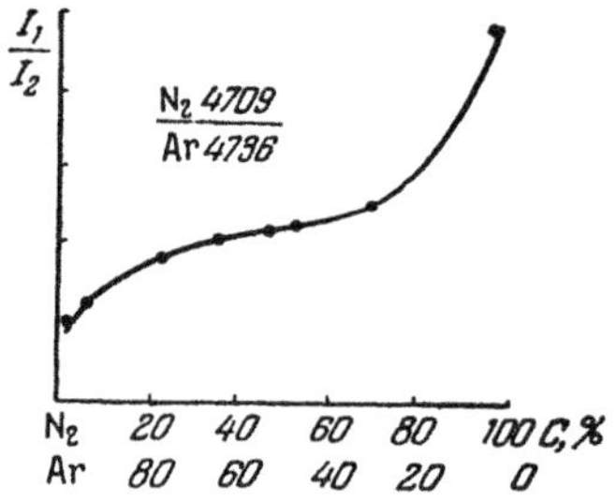

FIG. 73. Calibration curve for analyzing argon-nitrogen mixtures.

Traces of argon in nitrogen can also be detected if the mixture is excited in a high frequency discharge, using the positive column region. For high analytical sensitivity, it is necessary to use small-diameter capillaries (0.4 mm) and low pressures (0.2 mm Hg). In a small-bore capillary the nitrogen molecule dissociates into atoms, and a line spectrum is excited as a result. The calibration curve for argon concentrations in nitrogen ranging from 0.2 to 1.5% is shown in Fig. 74. For a single determination, the standard deviation is 6%. The intensity of the glow decreases sharply at low pressures. Thus, to increase the intensity of light falling upon the spectrograph slit, the discharge tube was not set up parallel to the slit, but was mounted at right angles to it (the end facing the slit). The tube end consisted of a flat quartz window, and a sharply enlarged image of the capillary was projected upon the slit. This permitted reduction of the exposure to 1/3 or 1/4 of what would have been otherwise necessary.

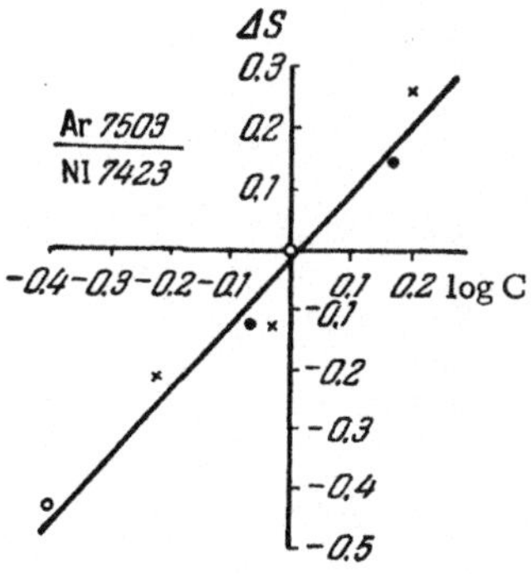

FIG. 74. Calibration curve for determining argon (concentrations ranging from 0.2 to 1.5%) in argon-nitrogen mixtures.

The nature of the glow produced by a high-frequency discharge is similar to the glow in a hollow cathode [367]. To observe this high frequency discharge we used a 12 mm diameter tube (Fig. 25a). The spectra were photographed in the middle and lower portions of the tube (as indicated by the arrows). When photographing the lower part of the tube, the light falling on the spectrograph came from the portion of the tube enclosed by the external electrode. The spectra from the middle part differed markedly from those from the lower part of the tube. The former were identical with the normal positive column spectra, while the glow in the lower section of the tube produced spectra similar to those obtained in hollow-cathode glow. Thus, in the case of nitrogen, molecular spectra predominated in the middle, while atomic spectra were most frequent in the lower part. The high frequency "hollow cathode" can be used in determining traces of a difficult-to-excite component in an easily excited primary constituent. Figure 75 shows the calibration curve for traces of argon in nitrogen. The curve was plotted at $p = 6 \cdot 10^{-2}$ mm Hg and $i = 300$ mA. The "hollow cathode" diameter was 12 mm. The slope of this calibration curve is very small, but it is nevertheless convenient to work with a large-diameter tube.

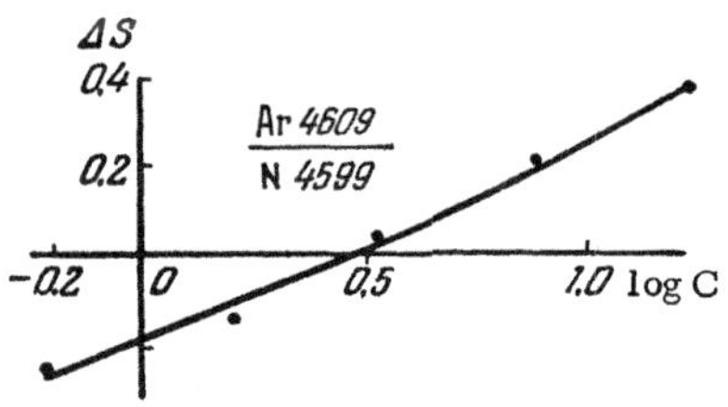

FIG. 75. Calibration curve for determining traces of argon in argon–nitrogen mixtures.

Analysis of argon in air was described in [361, 366, 382 and 415]. The results of these experiments are described in Section 25.

Krypton determination in oxygen

The high frequency discharge is also suitable for this analysis. While developing methods for krypton determination in oxygen, we noticed that the first few photographs taken after the sample had been admitted to the discharge tube showed krypton lines of far greater intensity than any of the subsequent photographs. Thus, the difference in optical density (blackening) ΔS for the analytical pair Krλ 5870 Å–OIλ 5437 Å in three consecutive photographs at 1.5% krypton in oxygen was 0.28, 0.11 and 0.12, respectively. The respective values for 2.3% krypton in oxygen were 0.55, 0.34 and 0.32. The second and third photographs gave results of excellent reproducibility, while the first photos yielded values with large scatter. It follows that some time should be allowed to elapse after admission of the sample to the discharge tube before photography is started. In the above case the discharge should be maintained for not less than 5 min. An identical effect is observed in the analysis for low nitrogen concentrations in argon.

The calibration curve for determining traces of krypton in oxygen is shown in Fig. 76. The photographs were obtained at p = 0.60 mm Hg, i = 350 mA, capillary bore of 0.5 mm. The analysis was based on the atomic oxygen line, since the emission spectrum of a neutral oxygen molecule is extremely difficult to obtain.

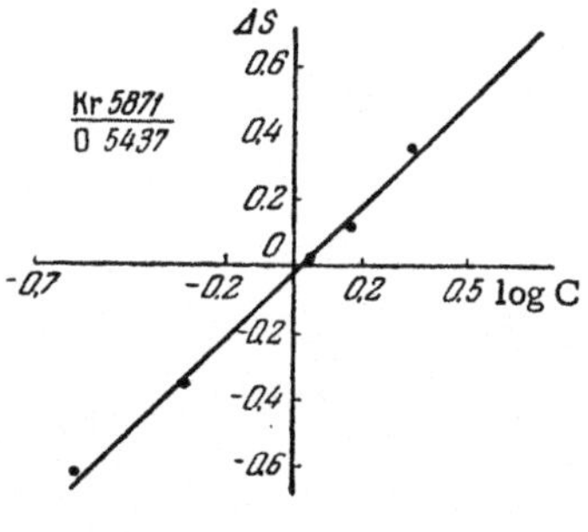

FIG. 76. Calibration curve for determining krypton concentrations in a krypton–oxygen mixture.

Helium determination in hydrogen

The limit of detection of helium in hydrogen in a high frequency discharge produced in a 0.5 mm I.D. capillary under a pressure of 1 mm Hg does not exceed 0.1%. Alongside the atomic hydrogen and helium lines, the spectrum shows a highly intense spectrum of the H_2 molecule. A suitable pair of analytical lines is Heλ 4471 Å—H_γ λ

4340 Å. The slope of the calibration curve is flat over the low concentration section. The mean error of the analysis can be as high as 8%.

IV. Analyzing mixtures of two molecular gases

Emission spectra from a mixture of molecular gases produced by conventional light sources as a rule do not correspond at all to the composition of the sample. The electric discharge causes dissociation of the majority of molecules into atoms or free radicals, with the result that even a single molecular gas will actually contain two components. Thus, the spectrum shows lines of atoms as well as molecular bands corresponding to excited molecules or free radicals. In addition, the discharge may produce new compounds in the mixture; their spectra may then also appear [367, 425].

All these factors greatly complicate the quantitative analysis of molecular gas mixtures by means of emission spectra. Thus, it is not possible to differentiate between CO and CO_2, H_2O and H_2, to separate the individual gaseous hydrocarbons, etc. Nevertheless, aside from free radicals which may be present in the sample, emission spectroscopy is capable of determining the components of a number of mixtures consisting of molecular gases with fairly high accuracy.

Determination of nitrogen in carbon dioxide

As an example, we shall discuss the analysis of a nitrogen-carbon dioxide mixture. Its high-frequency discharge spectrum will display N_2, CO and CN bands. The CO is the result of dissociation of the CO_2, while the CN is produced on reactions of CO with N_2. An equilibrium is established a few minutes after the start of the discharge. The nitrogen content in the carbon dioxide can be determined from both the nitrogen and the CN bands. Calibration curves are plotted from relative intensities of the molecular head band. The width of the spectrograph slit must be kept strictly constant throughout the calibration and analysis, since the head band intensity is a function of this width (one records a group of overlapping lines in the rotational structure of the band).* Calibration curves for concentrations ranging from 0.5 to 25% are shown in Fig. 77. The spectra were recorded at p = 3.5 mm Hg,** i = 350 mA, d = 5 mm, exposure 2-5 min [367]. An alternative illustration of analytical procedures used with mixtures of molecular gases is provided by the analysis of air samples taken near the ground (see Section 25).

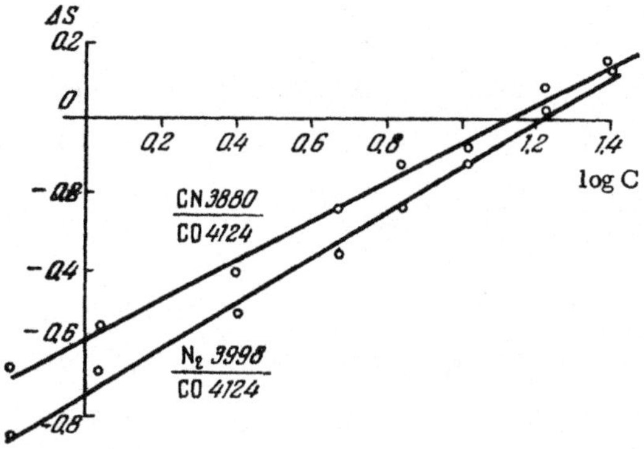

FIG. 77. Calibration curve for determining nitrogen in a nitrogen-carbon dioxide mixture.

24. QUANTITATIVE ANALYSIS OF MULTICOMPONENT MIXTURES

Analysis of multicomponent mixtures inevitably involves the effect of the third and subsequent components. The third component effect is especially pronounced in cases where critical potentials of the minor constituents are lower than that of the primary component.

*It follows that in cases where the lines of the rotational structure are sufficiently intense for photometry, the analyst should use one of the components of the rotational structure rather than the head band.

**The pressure in this case was selected on the basis of incidental considerations. A higher sensitivity of analysis could have been attained at greater pressures.

Such effects are also involved in analysis of alloys, ores and minerals, and were pointed out time and again by various researchers. Spectral analysis of gases is usually more affected by the third component factor than analysis of alloys and minerals which uses a spark or an arc discharge. In the case of gas samples, the third component effect may 1) alter the discharge conditions (i.e., produce an electron temperature drop) due to the presence of additional elements with low ionization potentials; 2) afford conditions conducive to chemical reactions in the discharge; 3) produce conditions favorable to collisions of the second kind.

Several methods have been developed for analyzing multicomponent mixtures. One such procedure, designed for analysis of ternary mixtures, may be used only when the line intensities are affected by only one of the components [371]. An example of this method is the case where helium is the primary component, while neon and argon, in concentrations ranging from 0.8 to 7%, are the minor constituents. The addition of neon does not appreciably alter the ratio of line intensities of argon and helium, whereas the addition of argon causes a change in the corresponding ratio for helium and neon. This is because the critical potentials of helium and neon differ far less than do the critical potentials of helium and argon, and thus the addition of neon to helium has little effect on the electron temperature (this temperature is already sharply reduced by the presence of argon in the mixture).

In the actual analytical procedure, a calibration curve was established for determination of argon in helium (see Fig. 78a). In constructing the curve, the absence of a neon effect on the ratio of line intensities of argon and helium was verified. The calibration curve was plotted from photographed spectra taken under the following conditions: a high frequency discharge; p = 1.4 mm Hg; i = 300 mA; tube diameter = 5 mm. Next, the calibration curves for neon in an argon-neon-helium mixture were plotted under the same conditions but at varying argon concentrations (see Fig. 78b) [371]. Thus, the method boils down to the following: the argon concentration is determined from the first calibration curve, after which one knows which curve from the second family of graphs in Fig. 78b can be used to determine the concentration of neon.

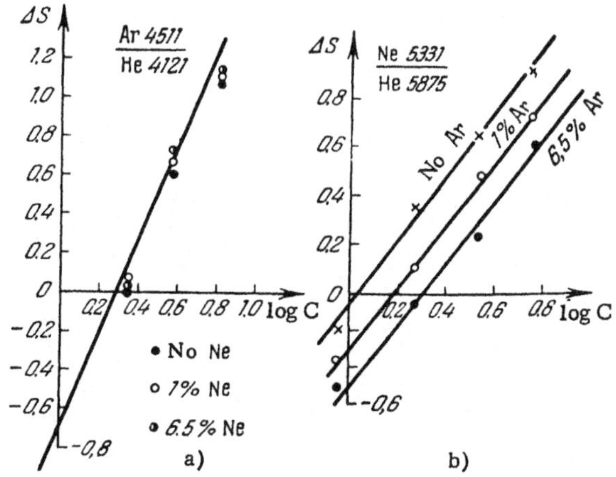

FIG. 78. Calibration curves for determining argon (a) and neon (b) concentrations in argon-neon-helium.

The method of plotting a family of curves was also used to determine argon in air when argon and oxygen concentrations were variable [382] (see Section 25). It was possible to analyze the mixture because changes in argon concentration did not affect the oxygen determination, and a change in oxygen concentration caused only a parallel displacement of the calibration curves for the argon-nitrogen mixture.

However, the third component effect does not always produce a mere parallel shift of the calibration curves: sometimes, it also causes a change in the slope of the curves. Thus, Krasnova and Schreyder observed such changes while developing an analytical procedure for the determination of traces of oxygen and nitrogen in hydrogen. The discharge was excited by means of an h-f oscillator in a 5 mm capillary under a pressure of 1 mm Hg. The addition of oxygen (see Fig. 79a) did not shift the calibration curve for nitrogen in hydrogen. The addition of nitrogen, on the other hand, caused not only a shift of the calibration curves but also a change of slope (see Fig. 79b).

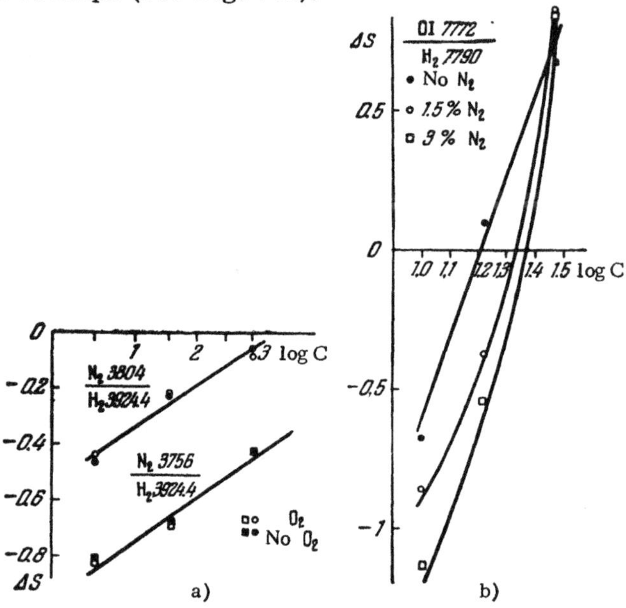

FIG. 79. Calibration curves for determining nitrogen (a) and oxygen (b) in an oxygen-nitrogen-hydrogen mixture.

As we stated earlier, this method for analyzing ternary mixtures has a limited application. In cases where both minor components affect each other, one should select another gas as the reference element. The line intensities of that gas should not vary with changes in concentration of the other components. Duffendack and Wolfe [381] used this technique in analyzing mixtures of nitrogen, carbon monoxide, hydrogen and oxygen. The discharge tube (a quartz capillary 3 mm I.D.) had internal nickel electrodes and the discharge current was 25 mA. To reduce the adsorption of gas during the discharge, large amounts of helium were introduced (the helium pressure varied between 2 and 8 mm Hg in different experiments). Helium was used because it is adsorbed on the electrodes and walls of the tube to a lesser extent than the other gases, and its ionization potential is also higher. Hence, the addition of helium does not reduce the electron velocity. However, helium could not be used as an internal standard, since its line intensities vary sharply with the concentration of the other components. The best internal standard proved to be argon, which was used at partial pressures ranging from 0.01 to 0.07 mm Hg. (Naturally, the immunity of the argon line to other components was prechecked.) The argon-helium mixture was admitted to the thoroughly evacuated discharge tube through a trap cooled with liquid air. The purity of the mixture was checked by photographic recording of the discharge. The first photographs displayed some impurity lines, but as the discharge continued these lines vanished.

To obtain a plot of the analytical curves, known quantities of hydrogen (partial pressures of 0.0001 to 0.0075 mm Hg), nitrogen (0.004 to 0.25 mm Hg), oxygen (0.002 to 0.12 mm Hg) and carbon monoxide (0.001 to 0.10 mm Hg) were introduced into the discharge tube. The calibration curve for small nitrogen concentrations in helium is shown in Fig. 80. Similar curves were obtained for hydrogen, oxygen and carbon monoxide in helium.

To determine the composition of an unknown mixture, an accurately known quantity of the sample was introduced into the discharge tube containing a helium-argon mixture (of the same composition as used in plotting the calibration curve) and the spectrum was photographed. If the helium-argon spectrum showed no additional lines or if such lines were of low intensity, another portion of the unknown mixture was admitted the tube, and the spectrum was rephotographed. The sequence was repeated until all the additional elements were determined from the calibration curves. The composition of the unknown mixture could then be determined using the known ratio of partial pressures of helium and the unknown sample.

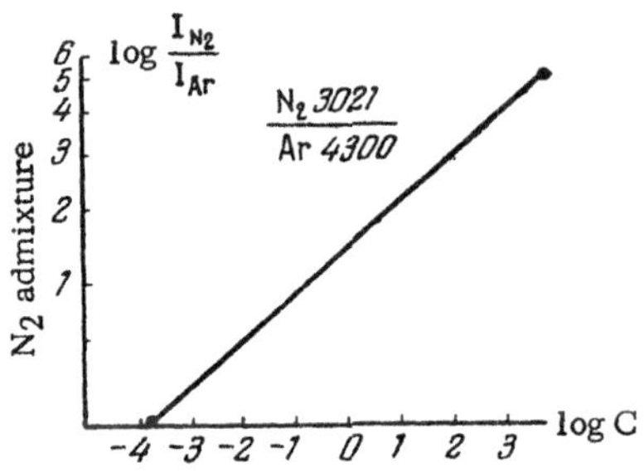

FIG. 80. Calibration curve for determination of low concentration of nitrogen in helium.

Helium dilution is also useful in microanalysis of air because such dilution circumvents the mutual interference of air components [419].

It would seem that if the primary component of a mixture is an inert gas with a high excitation potential and the minor components, which have lower excitation potentials, are present in relatively insignificant amounts, then the mutual interference of the components can be neglected, within the limits of experimental error. In this case, the analysis of a multicomponent mixture involves, in effect, nothing more than a simultaneous determination of several binary mixtures. The correctness of this hypothesis was confirmed by Bochkova and Chernysheva [384] in determining traces of nitrogen and hydrogen in helium. The mixtures were excited by means of a high frequency discharge in a 1-1.5 mm I.D. capillary, the hydrogen and nitrogen concentrations being in the 10^{-3} and 10^{-1} % range. The hydrogen was determined by means of the $H_\beta\lambda$ 4861 Å — Heλ 5047 Å line pair and the $N_2^+\lambda$ 4278 Å — Heλ 5047 Å pair was used for nitrogen. Line intensities were measured successively by means of an FEU-17 photomultiplier installed in back of the exit slit of an ISP-51 spectrograph (see Appendixes III and IV). The photomultiplier was powered by a stabilized rectifier, and the photocurrent was read off from a microammeter. The analytical procedure used three standards, with helium of spectral purity used in the preparation of standard mixtures. The residual nitrogen and hydrogen in helium was determined by the method of additions. In preparing the calibration curves, the ratios of line intensity for hydrogen-helium and nitrogen-helium mixtures (proportional to the respective photocurrent ratios) were plotted on the ordinate with the minor component concentrations on the abscissa. This procedure gave a series of linear calibration plots at a given pressure, each curve covering a range of minor component concentrations varying by a factor of 10. It was shown that if the nitrogen concentration does not exceed $2 \cdot 10^{-2}$%, this component does not affect the analysis of H_2 in He. When the nitrogen content is of the order of 0.1%, the error in determining hydrogen concentrations (in helium) of the order of 0.01% is within the limits of the accuracy of the procedure ($\sim$10%). However, the error in determining concentrations of the order of 0.001% hydrogen may be as high as 40–50%.

The limit of detection of hydrogen in helium is 10^{-3}% (at a pressure of 20 mm Hg). This limit is not, however, imposed by limitations of spectral analysis per se, but by the hydrogen purity attained in the tube. The evolution of hydrogen and water vapor from the tube walls during discharge reduces the analytical sensitivity for traces of hydrogen, thus adversely affecting the accuracy of the method.

In testing for the effect of hydrogen on the determination of nitrogen in a ternary helium-nitrogen-hydrogen mixture [384], it was found that the addiditon of 0.1% hydrogen to a helium-nitrogen mixture has no effect on detection of nitrogen concentrations of the order of 0.01 to 0.1%. In the nitrogen concentration range of 10^{-2}–10^{-3}%, hydrogen concentrations of the order of 10^{-2}% do not hinder the nitrogen detection. The limit of detection for nitrogen is 10^{-4}%. The percentages of nitrogen and hydrogen do not differ by more than an order of magnitude, and therefore the mixture can be analyzed simultaneously for both components. Under industrial conditions the analysis is most conveniently carried out in a stream of gas, using an arrangement of the type shown in Fig. 19. The time required for simultaneous analysis for both components does not exceed 5 minutes.

In some instances the third component effect can be eliminated by a trick, as in analysis of the He–Ne–Ar–Kr and He–Ne–Kr–Xe mixtures [383]. The method resorted to was the familiar spectroscopic technique of stabilizing the excitation conditions by introducing known quantities of a component whose excitation potential is lower than that of any of the other components. This technique, while useful on many occasions, is not altogether advantageous since it lowers the sensitivity.

The problem of analyzing multicomponent gas mixtures has been successfully solved in many cases. Some of these successes were described in this section, and others will be discussed in Sections 25 and 26.

25. ANALYSIS OF AIR

Since air is a multicomponent gas mixture, its quantitative analysis is difficult. Air consists of nitrogen and oxygen (basic constituents), as well as some hydrogen, carbon dioxide, water vapor and inert gases (total about 1%) of which argon is the most abundant (about 0.95%).

The composition of air near the ground is generally determined by chemical and physicochemical methods, which are fairly accurate in identifying the active components (O_2, N_2, H_2O, H_2, CO_2). The chemical procedures are, however, not suitable for analyzing inert gases. Fractional distillation with subsequent measurements of gas volumes is accurate only when relatively large gas samples are available [1]. The chemical and physicochemical methods become completely unsuitable for analyzing the composition of upper atmospheric air, a case where the samples are usually extremely small. In this case, optical spectroscopy and mass-spectrometry are the only available means of analysis.

Analysis of surface air. Determination of hydrogen and moisture in air

Hydrogen and water vapor in air are determined from the intensities of lines of atomic hydrogen (Balmer series). These lines are produced in the discharge tube by atomic or molecular hydrogen, as well as water vapor. If the water is frozen out, then one can determine the hydrogen content alone.

The direct analysis of hydrogen in air is capable of determining only concentrations not lower than ~1% [370, 403]. The limit of detection can be lowered by one order of magnitude by adding helium in quantities ranging from 10 to 50% of the sample [426].

Microspectroscopic analysis of traces of moisture in air at reduced and at normal pressures was described by Garger [427]. The spectrum was excited in a glow discharge produced in a Frisch all-purpose tube, using an induction coil generating 10–30 kV when combined with the Swenticki arc activator [315]. A condenser discharge can also be used, according to Garger. The discharge tube was filled with a mixture of air and water vapor. The partial pressure of the water was determined by measuring the temperature in a tube extension containing distilled water, which was placed in a Dewar flask filled with a cooling mixture. The H_α line was used for determining water in concentration of the order of 10^{-3} %. Moisture contents of the order of 0.01%-0.2% lead to the appearance of the H_β and H_γ lines as well as the OHλ 3064 Å band. Upon increasing the moisture content to 3.5%, the OIλ 6158 line of oxygen appeared and the nitrogen bands decreased in intensity. Garger developed a procedure for determining water in air both at reduced and at normal pressures. The limits of reliable measurements were 0.02-2%, with an error of 4-10%.

Determination of carbon dioxide and carbon monoxide

Both the carbon dioxide and the carbon monoxide present in the air are identified in the spectrum by means of the CO bands, since during the discharge carbon dioxide dissociates into carbon monoxide and oxygen. As a result, it is impossible to differentiate spectroscopically between CO and CO_2. Where each of these gases must be determined separately, spectroscopy is combined with some technique based on chemical absorption of one of the components. Occasionally simple freezing out of CO_2 will suffice, since it is virtually impossible to remove CO by this method.

The limit of detection of carbon dioxide and carbon monoxide in air, when a high frequency discharge is used, is about 1%. The analytical pairs are COλ 4835 Å—$N_2\lambda$ 4709 Å and COλ 5198 Å—$N_2^+\lambda$ 5228 Å; the analysis is carried out in a 0.5 mm I.D. capillary at a pressure of 1-2 mm Hg. The analytical error is of the order of 5%. In the concentration range of 1-30%, calibration curves are linear.

Wignen and Van-Tiggelen [425] determined carbon dioxide in air in which the oxygen concentration was variable. An r-f oscillator generating an 11 meter wave was used for excitation. The discharge was produced in a 5 mm I.D. glass Geissler tube at a pressure of

2.5 mm Hg. The sample-filled tubes were fused to the remainder of the system. These experiments showed that nitrogen is more intensively adsorbed on the discharge tube walls than carbon dioxide. To improve the reproducibility of the results, the spectra were photographed during the first 2-5 min after the discharge current had been switched on. The pressure in the discharge tube varied from 2 to 3.1 mm Hg, but this did not alter the relative intensities of the $N_2\lambda$ 3159.3 Å and COλ 3127 Å bands. Variations in the oxygen concentration also failed to alter the ratio of band intensities of N_2 to CO, provided the oxygen content was below 20%.

The procedure developed in [425] allows nitrogen determination over the 3.5-37.5% concentration range, and carbon dioxide determination at concentrations ranging from 46 to 95%. The mean error of analysis amounted to 3% in the case of nitrogen and 1% in the case of CO_2. The reported accuracy of the carbon dioxide determination appears doubtful, considering that during fusion of the glass tubes to the system large quantities of CO_2 must have evolved from the glass and must have contaminated the sample. Such sources of impurities are very difficult to control.

A very high analytical sensitivity for carbon monoxide in air is obtained in the procedure of [428], which is based on liberation of Hg in the reaction

$$CO + HgO \rightarrow Hg + CO_2.$$

The CO content was determined from the amount of liberated mercury, measured in terms of absorption of the mercury resonance line λ 2537 Å. The limit of detection for CO in this method is 10^{-6} %.

Determination of oxygen

The oxygen content in air is easily determined in the concentration range of 5 to 50%, using the line pair OIλ 7772 Å—NIλ 7468 Å.

The mixture is excited by a high frequency (6 Mc) discharge in a 1 mm I.D. capillary at pressures of a few mm Hg. Lower pressure in the discharge tube improves the analytical sensitivity. Hydrogen, water vapor and carbon dioxide affect the results. Water and carbon dioxide dissociate during the discharge, releasing additional oxygen. The presence of CO_2 in amounts below 1% was shown to have no effect on the oxygen analysis. When the CO_2 content is higher than 1%, the third component effect, which leads to a displacement of the calibration curves, must be taken into consideration.

Nitrogen and oxygen are intensively adsorbed on the tube walls during the discharge. As a result, the ratio of line intensities of oxygen and nitrogen varies with time. To achieve good reproducibility, this ratio should always be recorded at some fixed time after the start of the discharge. As the oxygen concentration is increased, the intensity of both the molecular bands and atomic lines of nitrogen decreases. As a result, the calibration curves invariably have slopes greater than unity. Future methods for checking rapid changes in the O_2 content of the air can be based on these intensity variations of nitrogen bands. In addition, the nitrogen radiation may be separated for photoelectric recording by means of a filter with a fairly wide passband in the 3900-4200 Å region, thus affording the basis for a compact, simple design.

Ozone determination in air is described in Chapter VI.

Analysis of inert gases in air. Determination of argon

The air contains relatively large amounts of argon (up to 1%). Therefore argon can be readily detected by any type of discharge: in a hollow cathode [366], a high frequency hollow cathode [367], a pulse discharge [371], the positive column of a high frequency discharge [362, 382], and a high-voltage spark discharge at pressures close to the atmospheric [415]. The smallest argon concentration in air which is detectable by high frequency discharge without preliminary enrichment of the sample is 0.1%.

Analyses of argon in air are affected by the oxygen concentration. As shown in [382], this effect is negligible when such concentrations do not exceed 10%. However, higher concentrations cause a parallel displacement of the calibration curves for the argon-nitrogen mixture (see Section 23). The experiments were carried out in a high frequency discharge, using a 0.5 mm I.D. capillary and a pressure of several mm Hg. The accuracy, based on the analytical pair Arλ 7504 Å — NIλ 7468 Å, is of the order of 5%.

A rapid and accurate procedure (taking less than 1 min) for argon determination in air, with photoelectric readout and recording of the results, was developed by Lee [361] (see Fig. 81). A high frequency oscillator was used to excite the discharge in a capillary. A continuous air stream was drawn through the discharge tube by means of a circulating system similar to the one proposed by Broida and Moyer (see Section 29). The discharge tube capillary was cooled with water. A photomultiplier was used to detect the radiation from argon line λ 4158 Å and nitrogen line λ 3998 Å. This radiation was separated by a monochromator with a diffraction grating.

In the concentration range of 0.9 to 1.1%, argon can be determined with high accuracy by measuring either the ratio of line intensities of argon and nitrogen or the absolute line intensities of argon. Lee has thoroughly explored the factors responsible for deviations of the ratio of the two signals (argon and nitrogen) from normal, and selected optimal excitation conditions and gas pressures in the discharge tube. In analyses based on the absolute intensities and on the ratio of intensities of the argon and nitrogen

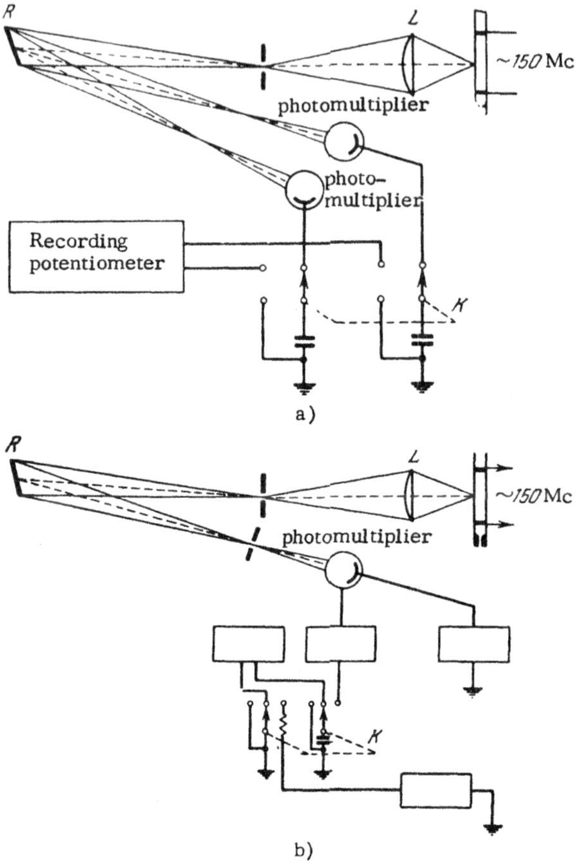

FIG. 81. Schematic diagram of the Lee unit for the determination of argon in air: a—procedure based on relative intensities of the argon and nitrogen lines; b—procedure based on the absolute intensities of argon lines.

lines, the analytical accuracy was as high as 95 ±0.43%* and 95 ± 1.4%, respectively. Analysis based on absolute measurements thus appears to be the more accurate. This is because the intensities of the nitrogen lines vary markedly with the discharge conditions, while the intensities of the argon lines remain constant. In such cases, there seems to be no point in relying on relative

*This means that 95% of the measurements involved errors not exceeding 0.43%.

measurements. According to Lee the analytical accuracy is improved by using a 2450 Mc oscillator. Since the concentration ratio of nitrogen to oxygen remained constant in these experiments, Lee did not need to consider possible changes in nitrogen and argon line intensities at varying oxygen concentrations.

Determination of helium

The helium content of air is very small. It cannot be detected spectroscopically even after all the active gases have been removed and the inert gas residue is analyzed. Helium concentration in a natural inert gas is of the order of $5 \cdot 10^{-4}$%, while its limit of detection in argon is no higher than 0.05-0.1%. It follows that the helium content in air cannot be determined spectroscopically unless the argon is removed. In some gases escaping from oil wells the helium content is higher, and can be determined without freeing the gases of argon.

The least helium concentration detectable in a high frequency discharge without freeing the air of active gases is of the order of 0.1%. The mixture is excited in a 0.5 mm I.D. capillary at pressures of 1-2 mm Hg, using the analytical pair Heλ 5876 Å—Nλ 5800 Å. At low concentrations, the calibration curve is not linear. The two factors responsible for the change of slope reduced relative sensitivity, and the presence of a background arising from molecular nitrogen bands. Helium can be determined in air in a pulse discharge [271]. The limit of detection is 0.05% at the following discharge parameters: tube I.D. = 15 mm; pressure = 8 mm Hg; voltage = 3000 V; C = 12 microfarads. The system had no discharge gap, and the discharge was fired by means of a Tesla transformer. The spectrum was photographed over an interval of a few pulses, without time scanning. Isolation of a specified time interval corresponding to a proper pulse should considerably increase the intensity of the helium lines with respect to the background, and will thus improve the analytical sensitivity.

Gladushchak and Schreider used a discharge inside a pulsed hollow cathode to determine low helium concentrations in air. A bank of condensers, capacity 5 microfarads, potential 10 kV, was discharged by thyratron switches and thereby sent a current flowing through the discharge tube. The hollow cathode (diameter—10 mm) was made of aluminum, a metal that has no great tendency to sputter, so that the discharge largely took place in the sample, rather than in a metal vapor. The tube pressure was of the order of 0.5 mm Hg. The limit of detection was about 0.02%.

Pulse sources appear to ensure a higher relative sensitivity in analysis of helium in air than high frequency discharges. One reason for this is that the end helium line λ 5876 Å superposes on nitrogen bands, which are strongly excited in a high frequency discharge, but fail to do so in a pulsed discharge.

Determination of krypton

The krypton of the air can be detected spectroscopically in any concentration up to the normal one provided all active impurities have been removed, leaving a residue of inert gases. The smallest air sample sufficient for a krypton determination is ~ 10 cm^3. The initial purification (lasting 5 min) is carried out in an arc using calcium electrodes. The complete removal of the residual nitrogen and oxygen is effected in Na or K vapor, while mainitaining the discharge in the tube over a period of 20 min. Following such purification the high-frequency discharge spectrum will show nothing but an Ar line and the line Kr 4319 Å, on which determination is based. Our analysis showed $1.5 \cdot 10^{-4}\%$ in air, which, within the limits of experimental error, agrees well with other data. Higher Kr concentrations in air (above 0.2%) are easily detected without a preliminary purification.

Analysis of air samples from the upper atmosphere

The upper atmosphere air can be analyzed for only three components—nitrogen, oxygen and argon. The analysis differs little from that for surface air. The difficulty lies in the fact that the gases in the sample are easily adsorbed in the discharge tube, and only minute samples are available. To obtain reliable analytic results, two factors must be given careful consideration: 1) sampling of the air and storage of the sample; 2) analysis of microquantities of the gas.

Sampling and storage of air samples are described in detail by Mirtov [389]. Although his studies were of a rather specific nature, much of what had been found applies generally to the preliminary steps necessary for any gas microanalysis.

The procedure for determining oxygen, nitrogen and argon admixtures in microvolumes of air is described in [382]. The analysis used a unit designed for microdeterminations of gases (see Section 19). The microsample (from a cylinder containing all the available gas) was drawn by suction into a 200-250 cm^3 vessel and was compressed into a 15 cm long, 0.5 mm I.D. capillary. The glow was excited by means of a high frequency oscillator. The line pairs OIλ 7772 Å — NIλ 7468 Å and ArIλ 7503 Å — NIλ 7468 Å were used for oxygen and argon, respectively. Calibration curves for oxygen and argon ions in air are shown in Fig. 82. Changes in argon concentration do not alter the ratio of intensities of lines O and N and therefore do not affect the shape of the calibration curves. Changes in the oxygen concentration, on the other hand, alter the ratio for Ar and N lines, causing a parallel displacement of the calibration curves, a factor that must be reckoned with in analysis.

The mean error in determining oxygen was 15%, while for argon it was 8%. These large errors proved due to the averaging of results obtained in analyzing the air samples in a "weak" and a strong discharge (see Section 5). The actual error does not exceed 3.5-5% in the case of argon and 5-8% in the case of oxygen.

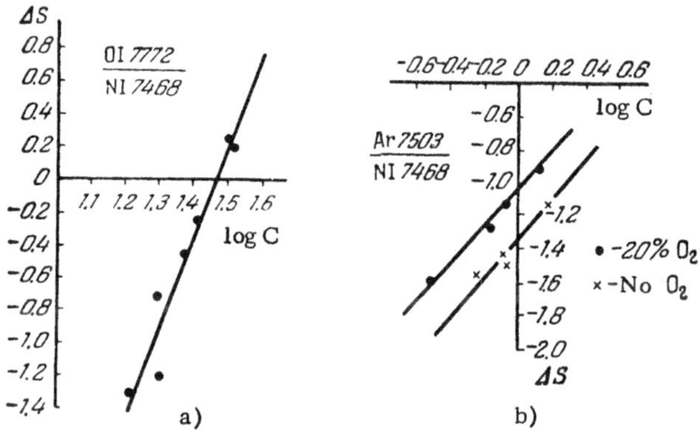

FIG. 82. Calibration curves for analysis of oxygen (a) and argon (b) in air without diluting the mixture with helium.

These errors are considerably reduced with larger gas samples. Mirtov [389] analyzed 0.5 liter samples of the upper atmosphere at $p = 10^{-3}$ mm Hg. The analytical error averaged 5% for oxygen and 3% for argon.

An alternative technique useful in analyzing minute air samples consists in adding an inert gas to the original mixture. The most suitable gas—which is a discharge carrier—is helium. It has the highest excitation potential of all the gases present, and consequently the latter are more readily excited than helium; for this reason, their detection limits in helium are 10^{-3}-10^{-5}%. In addition, the total gas mass of the sample is increased by the helium. This reduces the adverse effects associated with changes in the composition of the mixture, as well as the sorption and desorption of gas by the discharge tube walls.

To improve the reproducibility of results, the pressure of the helium added to the original air sample ($p = 1 \cdot 10^{-4}$ mm Hg, $V = 250$ cm^3) must exceed the pressure of the sample by a factor of more than 3 and not more than 100. The best results were obtained on adding a fivefold excess of helium. The addition of helium in 100-fold (or still greater) excesses allows quantitative spectroscopy of gas mixtures in volumes that could scarcely be analyzed otherwise. The minimum air sample required for a single-component determination can be reduced to $3 \cdot 10^{-5}$ mm Hg in a volume of 250 cm^3 (i.e., 0.01 mm^3 at STP).

Figure 83 shows calibration curves for analysis of argon and oxygen in air, obtained on diluting the original air mixture with an 80-fold amount of helium. Atomic nitrogen as well as helium lines can be used as the reference lines in the case of oxygen. The error in quantitative oxygen analysis is 10–12%. It should be pointed out that when helium is added to the mixture, changes in the oxygen concentration cease to cause shifts in the calibration curves used in analysis for argon. The absolute sensitivity of Ar analysis in air amounts to $5 \cdot 10^{-5}$ mm^3 at STP.

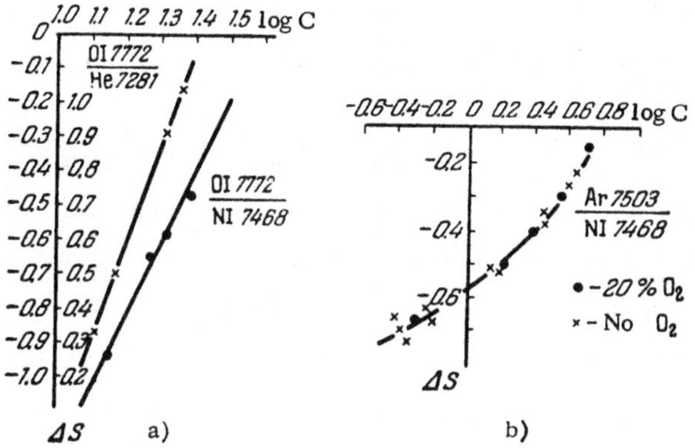

FIG. 83. Calibration curves for analysis of oxygen (a) and argon (b) in air following dilution of the mixture with helium. The negative ΔS values relate to the helium line, and the positive values to the nitrogen line.

The ionic and isotopic composition of upper atmosphere air may be analyzed with the aid of a mass spectrometer. Quantitative analytical data on the composition of the upper air are given in [389, 429-432].

26. FAST ANALYSIS OF GAS MIXTURES

In the preceding sections we have discussed various methods of quantitative spectral analysis where spectrographs or monochromators were used as dispersing systems. Since all such procedures involve a vacuum unit and complex spectroscopic equipment, most techniques designed for spectral gas analysis are too cumbersome for use outside a large laboratory. However, there is a great need for simple and rapid methods of analysis of gaseous media. Such procedures must be adaptable to use by semi-skilled personnel under plant conditions, and should ensure a sufficiently high accuracy and sensitivity of analysis. The so-called "fast" methods of spectral gas analysis meet such requirements. In many cases, especially where binary gas mixtures are analyzed, complex spectroscopic equipment can be replaced by a suitable monochromatic filter [417, 433]. This technique, widely practiced in absorption spectroscopy (see Chapter IV), is now being occasionally used in emission spectral analysis of metals. If a continuous gas flow is available, the vacuum unit can be considerably simplified [416]. Separation of radiation of desired wavelength by means of monochromatic filters, in turn, allows the use of simpler photoelectric units because the relative intensity of the light flux becomes sufficiently high [358, 359, 418]. In theory, most procedures of quantitative spectral analysis of gases can be simplified through such modifications.

Let us now review in some detail the few reported attempts at developing such methods. Servigne, de Montgareuil and Dominé

[417] worked out a simplified technique for analysis of nitrogen in argon and neon. The unit included a magnetron oscillator (f = 2450 Mc), an interference filter (λ_m= 3998 Å) and a vacuum photocell. The limit of detection was 10^{-4}%, gas consumption, 100 cm^3, and the analysis was finished within a few minutes. According to the authors, similar methods can be used for hydrogen in inert gases, whereby the line H_β (λ 4861 Å) is used.

The same method and equipment were used by Vernotte [312].

Simplified procedures for determination of nitrogen in argon and other inert gases, based on use of interference filters, are described by Bochkova, Razumovskaya, Frisch, Chernysheva and Sagaydak [418-420]. A vacuum unit such as shown in Fig. 19 was used, and the analysis was of the flow type. The gas is drawn by means of a forepump from the distributing manifold and through the discharge tube capillary. The pressure in the tube is regulated by means of valves and a manometer. The glow is excited by an r-f oscillator. The discharge radiation is projected by a condenser lens upon the photocathode of the photomultiplier FEU-19 (see Appendix IV) and is indicated (without amplification) by a microammeter. A glass light filter, with a transmission peak of about 3700 Å and $\Delta\lambda \approx$ 400 Å, is employed to separate the nitrogen bands in the λ 3600 Å region.

A working calibration chart is prepared on the basis of known standard mixtures, with nitrogen concentrations plotted along the abscissa and microammeter readings α, along the ordinates. The values of photocurrents α are proportional to the light flux of the radiation emitted by nitrogen bands, and isolated by means of the light filter. The light flux ratio α/α_0 can also be used, where α_0 is the photocurrent produced by the total radiation from the discharge tube. Since the nitrogen band intensity in argon is strongly dependent on the total pressure, it is essential that the pressure in the discharge tube be optimum (see Fig. 84) for every range of nitrogen concentrations in argon.

We shall next describe a method for determining the nitrogen content of argon of various purities.

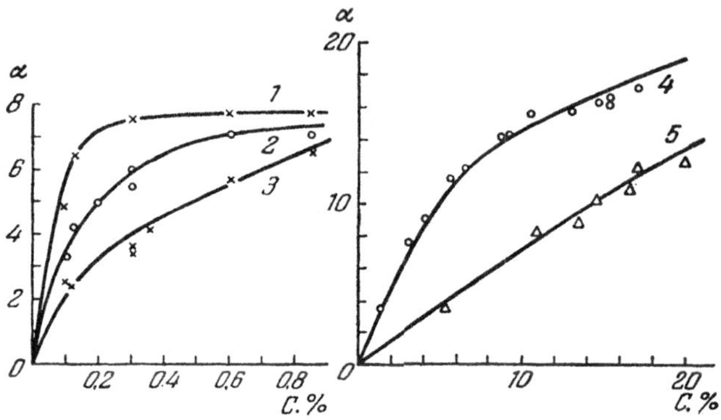

FIG. 84. Effect of pressure on the shape of calibration curves. Nitrogen determination in argon (capillary 1.2 mm I.D.) 1—p = 100 mm Hg; 2—p = 80 mm Hg; 3—p = 50 mm Hg; 4—p = 10 mm Hg; 5—p = 3 mm Hg.

Technical-grade argon

Technical-grade argon contains from 9 to 15% nitrogen. At these nitrogen concentrations, high intensity molecular bands appear already at discharge tube pressures of 1-3 mm Hg. This pressure corresponds to curve 5 in Fig. 84; the curve is linear for all nitrogen concentrations below 20%, but has a relatively shallow slope. As the pressure is increased to 10 mm Hg the curve becomes nonlinear; its slope is high in the nitrogen concentration range of 1-10% and then markedly decreases in the 10-20% N_2 range. This curve may be used for analyzing nitrogen concentrations in the range of 1-10%. Above 10%, nitrogen must be determined at a lower pressure.

Pure and spectrally pure argon

The nitrogen content of pure argon should not, as a rule, exceed 0.5%. A discharge tube pressure of about 50-80 mm Hg is then required for the excitation of strong nitrogen bands. The so-called spectrally pure argon should contain less than 0.01% nitrogen. To determine such nitrogen traces, the pressure in the discharge tube must be several hundred mm Hg.

Linear calibration curves can virtually always be obtained for nitrogen in argon concentrations ranging from 0.001-0.01, 0.01-0.1, and 0.1-1%. This is done by selecting a discharge tube of proper diameter, and adjusting its pressure level. The curves can be plotted in terms of either absolute values of photocurrent α, or of the relative values α / α_0. Relative measurements can be made with the aid of a fixed calibration graph which remains constant over a long time. In absolute measurements, however, the calibration curves may undergo a parallel shift, due to instability of the unit. If the relative analysis uses a single photomultiplier, two readings must be taken—one with the filter, the other without—for determining α / α_0. An alternative, more convenient method is to obtain the α / α_0 ratio directly with the aid of a ratiometer circuit.

Various modifications of the photoelectric circuit for indicating the ratio of two luminous fluxes were discussed in Section 14.

Nitrogen determinations in argon may be seriously distorted by the presence of other gaseous impurities such as oxygen or carbon dioxide. Oxygen concentrations of the order of 0.1% do not affect the analysis of pure and technical-grade argon. At oxygen concentration of the order of 1%, a parallel shift of the curves is noted, producing results which are too low. In view of this, the unit has facilities (see Fig. 19) for scrubbing the sample free of oxygen. This is a trap with copper filings placed in a furnace at 350-400°C. In low concentrations carbon dioxide also does not affect nitrogen determination and is, furthermore, easily removed from argon.

Borok and Aleksandrov [358] separated the desired radiation by means of filters and designed an automatic gas analyzer for continuous nitrogen analysis in argon. They used two filters:

Table 2 Basic data yielded by simplified analysis

Gas mixture	Component determined	Concentration range, %	Excitation conditions	Filters	Radiation detector	Remarks
$Ar + N_2 + CO_2 + O_2 +$ hydrocarbons	N_2	10^{-3}–10^{-2} 10^{-2}–10^{-1} 10^{-1}–1	r-f discharge f = 6 Mc d = 1 mm p = 50-100 mm Hg gas stream	UFS-3 UFS-4	Photomultiplier FEU-17 Photomultiplier FEU-19	O_2 and CO_2 in concentrations lower than 0.1% do not affect the nitrogen analysis
		1-20	r-f discharge f = 6 Mc d = 1 mm p = 2-5 mm Hg no gas stream	λ_m = 366 millimicrons		
		0.1-1 0.01-0.1	r-f discharge f = 30 Mc d = 5 mm p = 0.3 mm Hg	S36-11 PS-11	STsB-3 STsB-6	[353]
		10^{-2}–10^{-1}	AC current f = 50 cps V = 2000 V d = 2 mm p = 10 mm Hg	λ_m = 390 millimicrons ZhS-11	Photomultiplier	[384]
		10^{-4} and higher	r-f discharge f = 2450 Mc p = 20 mm Hg r-f discharge f = 2450 Mc	λ_m = 400 millimicrons λ_m = 399.9 millimicrons	Photocell	[312] [417]
	Total hydrocarbon content	10^{-3}–10^{-2}	r-f discharge f = 6 Mc d = 1 mm p = 200 mm Hg	λ_m = 516.8 millimicrons	Photomultiplier FEU-19	

He + H_2 + N_2	N_2	10^{-3}–10^{-2} 10^{-2}–10^{-1}	r-f discharge f = 6 Mc d = 2 mm p = 40–60 mm Hg gas stream	λ_m = 420 millimicrons	Photomultiplier FEU-17	N_2 and H_2 are determined independently over the same 10:1 (upper to lower limit) concentration range
	H_2	10^{-3}–10^{-2} 10^{-2}–10^{-1}	r-f discharge f = 6 Mc d = 2 Mc p = 40–60 mm Hg	λ_m = 507 millimicrons λ_m = 486 millimicrons	Photomultiplier FEU-17 Photomultiplier FEU-19	
He + Ne	Ne	10^{-3}–1	r-f discharge f = 6 Mc d = 1 mm p = 100 mm Hg	λ_m = 640 millimicrons red filter	Photomultiplier FEU-19 Photomultiplier FEU-22 with amplifier	
He + Ne in air	Ne	20–35	r-f discharge f = 6 Mc d = 0.5 mm p = 2 mm Hg	λ_m = 640 millimicrons λ_m = 588 millimicrons	Photomultiplier FEU-22 Photomultiplier FEU-19 with amplifier	
	He	5–10	r-f discharge f = 6 Mc d = 0.5 mm p = 2 mm Hg	λ_m = 588 millimicrons λ_m = 447 millimicrons		

the first isolates primarily the 4000–5500 Å spectral region, which abounds in argon lines, while the transmission peak of the second filter coincides with the nitrogen bands 3600 Å. The use of a loop-shaped discharge tube (5 mm I.D.) allowed the use of photocells for indication. The effect of pressure on the shape of calibration curves was tested in special experiments. A working pressure of 0.3 mm Hg was selected on the basis of general considerations regarding the gas consumption, as well as the normal operating conditions of the oil diffusion pump producing the gas stream. The pressure is adjusted automatically. A detailed description of this gas analyzer and of its operation was given in Section 14.

Dianov-Klokov and Saltykova [359] used a filter with λ_m3990 Å, $\Delta\lambda$ 150 Å to separate the nitrogen radiation, plus a special filter for argon. The nitrogen in argon was again determined by the gas flow technique. The glow was excited in a discharge tube provided with internal electrodes (capillary 2 mm, I.D., pressure about 10 mm Hg). An original system operating with a single photomultiplier and a ratiometer (see Section 14) was used to indicate the intensity ratio of the two fluxes.

All these techniques, while embodying somewhat different designs, are based on the same principle of using light fluxes of high intensity. This allows considerable simplification of both the analytical procedure and the equipment. The general method is no less accurate than the more complex spectroscopic techniques, and at the same time it enables the analyst to carry out a determination within 2–3 min. It can be used to determine nitrogen in other inert gases, as well as hydrogen in helium, neon in helium, and a neon-helium mixture in nitrogen. All that is necessary is the selection of filters suitable to each analytical problem.

The results of the simplified methods are summarized in Table 2.

As was shown in the case of analysis of hydrogen in helium, the use of wideband filters reduces the relative sensitivity of the analysis.

27. GAS MIXTURE ANALYSIS IN GAS DISCHARGE TUBES

Pure inert gases, their mixtures, and hydrogen are used in industry to fill various types of gas-discharge devices (gas rectifiers, thyratrons, voltage stabilizers). The performance of such equipment, as well as its useful life, depends to a large extent on the composition of the filling gas, as well as maintenance of its purity during service. The filling gas cannot be sampled for spectroscopy while the equipment is in operation. The composition of the gas can only be checked either before the final assembly of the device, or when it is no longer in service.

In many cases, however, the composition and purity of the filling gases can be tested in the gas discharge device itself, during its normal operation. While the data on this are, so far, quite scarce, they are of great interest, inasmuch as they broaden the scope of procedures of emission analysis.

In one study [293] neon admixtures (10^{-3}–1%) in helium were analyzed directly in a TG-1P tube (filament voltage, 6.3 V; plate current, 35 mA; working gas pressure, 2 mm Hg) by the three-standard method. An arc discharge occurs in this tube during normal operation. The glow was observed through 1 mm diameter apertures in the tube plate. These performations did not affect the tube characteristics. The reference tubes were filled with the reference mixtures in a separate installation. Direct analysis in TG-1P tubes yielded results which showed good agreement with those obtained by the use of a high frequency discharge. No changes in the gas composition were noted at the end of the service life of the tubes. The same tubes, according to the authors of [293], can be used as light source in analyzing helium purity.

Kotlik and Lomonosova [294] recommend using a glow discharge produced in pilot lamp SN-1 for testing the composition of argon-helium and argon-neon mixtures. The lamp operated on an AC line current. The argon in helium analysis was based on the line pair Arλ 4159 Å—Heλ 4471 Å. In the argon concentration range of 0.5-2.0%, the operating pressure in the lamp was 30 mm Hg. In the argon in neon analysis, the pressure was 30 mm Hg for the 0.1-1% argon range and 20 mm Hg for the 2-8% argon range, whereby the line pair used was Arλ 4191 Å—Neλ 5341 Å. The spectra were recorded photographically. Photoelectric indication was the alternative method in argon-helium determinations. A λ_m 4200 Å interference filter was employed to separate the radiation due to argon lines.

Musson-Genon and Deschamps [434] determined the water vapor in tubes such as those used in radar installations while the tubes were in operation. They did this by monitoring changes in intensity of the OHλ 3064 Å band. The discharge was excited by means of a pulse high frequency oscillator at a tube pressure ranging from 1 to 20 mm Hg. The tube filling consisted of carbon dioxide (as the principal component), hydrogen and water vapor. The calibration curve, plotted prior to the operation on the basis of known water vapor concentrations (the water vapor pressure varied from 0.5 to 16 mm Hg), served to follow changes in these concentrations while the tubes were operating. The method is not without faults, since other factors, besides water vapor, may cause the appearance of OH bands in a spectrum.

It is still more difficult to make emission-spectroscopical analyses of residual gases as well as gases liberated during the normal operation of vacuum equipment. Such analytical tasks are generally carried out by use of mass spectrometer techniques.

A mass spectrometer of special design, devised for analyzing gases released during the manufacture and normal operation of electronic tubes, was described by Morrison [435]. The tubes are opened under vacuum, by means of a special device. The gases are collected in a measuring pipet with the aid of a small Töpler pump, and are then analyzed in the mass spectrometer. It was shown experimentally that the bulk of the gas consists of carbon monoxide and evolves from the plate and its supporting structure. Gases collected in this manner can also be analyzed by microspectroscopical microtechniques.

Residual gases can also be analyzed with the aid of standard resonance mass spectrometers of the omegatron type [436], at pressures as low as 10^{-11} mm Hg.

28. ANALYSIS OF GAS MIXTURES IN MEDICINE

The first experiments on determination of nitrogen in exhaled gases date back to 1944, when Lilly and Anderson [437] developed a simple gas analyzer. A tube operating on a stabilized direct current (2-5 mA) was the light source, and the glow of the positive column was analyzed. A filter and photocell were used to separate the wavelength range of 3100-4800 Å. It was established by special tests that the photocell failed to detect the radiation emitted by the O_2, CO_2 and water vapor impurities, hence it indicated changes in the nitrogen concentration. This gas analyzer was further improved [438] and could be used for nitrogen determinations in the concentration range of 0.5 to 80%. The analysis was of the flow type.

White et al. [439-443] carried out a series of studies on the analysis of exhaled gases. Tubes of the type used by Lilly and Anderson, as well as tubes using external electrodes, were the light sources. The oscillator frequency was varied between $3 \cdot 10^3$-$6 \cdot 10^3$ cycles. A modified Beckmann spectrophotometer, operating as monochromator, indicated the radiation. Photographic measurements were taken simultaneously. Pressure in the discharge tube was varied from a few tenths to several mm Hg. Analytical procedures were developed for CO_2, N_2 and O_2. In the early experiments, the CO_2 analysis was based on the band CO_2^+ λ 4815 Å, and in later studies, on CO_2^+ bands in the λ2883–2896 Å region. Calibration curves were plotted for CO_2^+ determination in the concentration range 3–10%. Nitrogen analysis, for concentrations ranging from 25–80%, was based on the λ2372 Å band. A triplet line λ7774 Å and a singlet line λ8446 Å were used to analyze oxygen in concentrations of 0.25–100%. The analysis was continuous, its total time not exceeding 0.1 sec.

Reviews of the studies on emission spectral analysis of exhaled gases were presented by White and Lovelage [438, 439].

Comparing the results obtained by mass spectrometric and spectroscopic nitrogen determinations, White [439] came to the conclusion that while spectroscopic analysis is less accurate than mass spectrometry, it is both more sensitive and less time-consuming.

Determination of changes in nitrogen content in medical research with the aid of a low frequency discharge was discussed by Nakagawa [444]. The analysis of a mixture of N_2, CO_2, H_2O and O_2 at rapidly changing oxygen content, based on the oxygen band absorption in the Schumann region, is described in Section 33.

29. SPECTRAL ANALYSIS OF ISOTOPES

Both atomic and molecular spectra can be used for determining the isotopic composition of a mixture.* Atoms of the same element that differ in weight (isotopes) have differing energy levels.

*The fundamentals and more important results of spectral analysis of isotopes are discussed in [209, 445 and 446].

This shift in levels leads to differences in the transition frequencies between levels, which in turn manifest themselves in altered wavelengths.

Two factors are responsible for isotopic shifts: 1) the nuclear mass is finite (it is not infinitely large with respect to the mass of an electron): 2) the nuclear volume is finite [19]. Differences between the masses of nuclei lead to changes in the Rydberg constant. The following equation holds for a single electron system:

$$\frac{\Delta\lambda}{\lambda} = \frac{m}{M}\left(\frac{1}{A_1} - \frac{1}{A_2}\right) = \frac{m}{M}\,\frac{(A_2 - A_1)}{A_1 A_2}, \tag{5.1}$$

where m is the electron mass, M is the proton mass, A_1 and A_2 denote the atomic weights of the two isotopes. It follows from (5.1) that the isotope shift is approximately inversely proportional to the square of the average atomic weight. In the case of light nuclei, the isotope shift which is caused by a change in the mass of the nucleus can be detected without using high resolution equipment. For a hydrogen-tritium mixture, the shift of the first few terms in the Balmer series is of the order of two angstroms (2.37 Å for H_α and 1.78 Å for H_β). A far less pronounced isotope shift is noted for helium lines, and a still smaller one for carbon and nitrogen lines. The elements from the middle of the periodic system are characterized by very small isotope shifts. In the case of the heavy elements the shift increases, as the volume effect becomes the primary factor. The nucleus can no longer be treated as a point, so that its dimensions must be taken into consideration [19].

Because of the low isotope shift values obtained for the atomic lines of nearly all elements, the determination must be based on molecular spectra, where the isotope shifts are generally considerably more pronounced. Such molecular shifts result from differences in the frequency of natural vibrations of the isotopic molecules, as well as from differences in their moments of inertia.

The energy of a molecule can be expressed as the sum of electronic (E_e), vibrational (E_v) and rotational (E_r) energies:

$$E = E_e + E_v + E_r.^* \tag{5.2}$$

The isotope mass has little effect on the electronic energy of a molecule, and hence the shift of energy levels, caused by shifts in the electronic levels, is neglected.

The isotope shifts will now be calculated for the vibrational levels. The vibrational energy is detemined from the equation

$$E_v = \hbar\omega\left(v + \frac{1}{2}\right) - \hbar x\omega\left(v + \frac{1}{2}\right)^2 + \ldots, \tag{5.3}$$

where $\hbar$ is Planck's constant divided by 2π, ω is the angular vibrational frequency, v is the vibrational quantum number, and x is the anharmonicity constant.

Neglecting the anharmonicity, we shall calculate the vibration frequency (ν_v) corresponding to a transition between two vibrational levels:

$$\nu_v = \omega\,(v' - v''). \tag{5.4}$$

The angular vibrational frequency ω can be found from the equation

$$\omega = \sqrt{\frac{k}{\mu}}, \tag{5.5}$$

where k is the quasielastic modulus ** and μ is reduced mass. For a diatomic molecule, we have

$$\mu = \frac{m_1 m_2}{m_1 + m_2}, \tag{5.6}$$

where m_1 and m_2 are the respective masses of the two nuclei α and β.

We thus obtain for the isotopic shift $\Delta\nu_v$

$$\nu_{v_2} - \nu_{v_1} = \Delta\nu_v = \frac{1}{2\pi}(\omega_2 - \omega_1)(v' - v''). \tag{5.7}$$

The values ω_1 and ω_2 will now be expressed in terms of the reduced

*Equation (5.2) is approximate, since to some extent these energies are interdependent.

**k is more commonly referred to as the force constant (Editor's note).

masses of the two isotope molecules:

$$\nu_{v_2} - \nu_{v_1} = \frac{\omega_1}{2\pi}(\rho - 1)(v' - v'') \tag{5.8}$$

where $\rho = \sqrt{\frac{\mu_2}{\mu_1}}$. Hence

$$\frac{\Delta\nu_v}{\nu_v} = \rho - 1. \tag{5.9}$$

Superimposed upon the vibrational isotopic effect is a rotational isotopic effect, which produces an additional shift of rotational lines [16]. The rotational energy E_r of a diatomic molecule is determined with the aid of the equation

$$E_r = hB_0J(J+1), \tag{5.10}$$

where h is the Planck constant, J is the rotational quantum number, B_0 is a constant whose value depends on the moment of inertia of the molecule, $I_0 = \mu r_0^2$, and r_0 is the distance between the nuclei:

$$B_0 = \frac{h}{8\pi^2 I_0}. \tag{5.11}$$

The expression for the vibration frequency corresponding to a transition between two rotational terms ($J-1 \rightarrow J$) can be written in the form

$$\nu_r = 2B_0J. \tag{5.12}$$

where J is the rotational quantum number of the upper state. Remembering equations (5.10), (5.11) and (5.12), we calculate $\Delta\nu_r/\nu_r$ for the two isotope molecules:

$$\frac{\Delta\nu_r}{\nu_r} = \rho^2 - 1. \tag{5.13}$$

The overall isotopic shift $\Delta\nu$ arising from the shift in both the vibrational and the rotational terms is given by the approximate expression

$$\Delta\nu = (\rho - 1)\nu_v + (\rho^2 - 1)\nu_r. \tag{5.14}$$

Since $\nu_v \gg \nu_r$, it follows that the isotope shift is determined largely by the shift of the vibrational levels.

We now turn to the discussion of the methods of isotope analysis, and the more important results obtained therefrom.

Thus far, analytical procedures have been developed for the gaseous mixtures of hydrogen isotopes [209, 290, 447-455], helium isotopes [209, 456, 457], carbon isotopes [209, 458, 459], and nitrogen isotopes [209, 460-464].

Spectroscopic studies of isotopic mixtures involve certain difficulties. Equipment of high resolving power is required for the resolution of spectral lines which differ in wavelength by only a few tenths of an angstrom, and for the quantitative measurement of the ratio of their intensities. Such high resolution equipment must be even more efficient in cases where the concentration ratio of one isotope to the other is very high, since the intensity measurement then involves a weak line located next to a very strong line. For this reason, procedures for the spectral analysis of isotopes use either diffraction grating spectrographs or Fabry-Perot standard interference filters in conjunction with a spectrograph or an appropriate light filter.

The analysis of isotopic mixtures, on the other hand, does not involve some difficulties encountered in analysis of ordinary gas mixtures which can be of an arbitrary composition. The requirements are not nearly as high with respect to the stability of the light source or the constancy of pressure in the discharge tube. Nor is it necessary to create discharge conditions especially favorable for the excitation of a particular component of the mixture. Any changes in discharge parameters should have the same effect on the lines of both isotopes, since the excitation potentials of isotopic spectral lines are very nearly the same, as are the effective cross sections of atomic excitations and the transition probabilities.

A basic problem in developing methods for analysis of isotopic mixtures concerns the feasibility of devising an analytical procedure which uses no reference standards. To start with, it is essential to check whether the changes in the ratio of isotopes in the mixture manifest themselves in changes in the ratio of the intensities of their spectral lines. If such a relationship does exist, there is no need for standards, and the intensity ratio of the isotopes can be taken as a measure of relative concentration. This intensity ratio is influenced by many factors. The most important of these, it would seem, is the separation of isotopes in the narrow capillaries of the input system, and of the discharge tube proper. It follows that analysis without standards requires conditions under which such separation does not occur [209]. The elimination of standards considerably simplifies and reduces the duration of the analytical procedure.

Let us now review in some detail the analyses of various isotopic mixtures.

Determining the isotopic composition of hydrogen

Most of the experimental work on isotope analysis was devoted to this problem. The early studies were aimed at evaluating the hydrogen enrichment with deuterium [401, 465]. The deuterium determination procedure was refined by Van Tiggelen [447], who carried out the analysis in a sealed discharge tube, using photographic recording of spectra. The earlier methods for analyzing mixtures of hydrogen isotopes were substantially modified by Broida et al. [448-450]. The analysis was of the flow type, with photoelectric indication of the spectra.

Broida and Moyer [448] showed that the ratio of line intensities of hydrogen and deuterium (which is the measure of the deuterium concentration) depends on the discharge tube diameter, the pressure in the tube, the current density, and the rate of gas flow through the discharge tube. The relative line intensities may also be affected by the presence of impurities. Figure 85 shows a schematic diagram of the Broida apparatus for determination of deuterium in hydrogen (deuterium concentration ranges from 76 to 95%). The hydrogen and deuterium were obtained electrolytically, and deuterium was freed of hydrogen by diffusion through a heated palladium capillary. Reference mixtures were prepared in a 3 liter vessel at a pressure of the order of 600 mm Hg. The analysis was of the flow type, to eliminate hydrogen adsorption on and its evolution from the discharge tube walls. The gas flow rate was regulated by using capillaries of various diameters. The mixture was excited in a 10 cm long, 4-8 mm I.D. discharge tube, connected to a 400 W high frequency oscillator. To achieve a more stable glow, the tube was cooled with running water. A diffraction grating monochromator (D = 10.4 Å/mm) coupled to a photomultiplier indicated the output radiation.

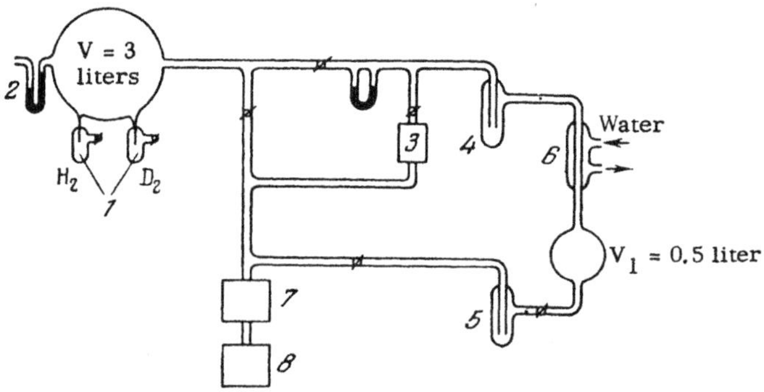

FIG. 85. Broida's arrangement for analyzing hydrogen-deuterium mixtures. 1—Palladium capillaries; 2—U-tube manometer; 3—McLeod gauge; 4 and 5—trap for freezing out mercury; 6—discharge tube; 7—high vacuum pump; 8—forepump.

The hydrogen:deuterium line intensity ratio varies with the pressure in the tube. To maintain a steady pressure, a 500 cm^3

vessel was connected to the tube. In addition, a Töpler pump was provided which made it possible to vary the pressure, increasing it approximately fourfold without changing the composition of the mixture. The authors of [448] point out that when one of the isotopes predominates in the mixture, the intensity of each line need not be measured with an accuracy of 0.5% in order to determine the concentration with this accuracy. The error in determining the ratio of concentrations of deuterium and hydrogen $\delta(D/H)$ is determined from the equation

$$\delta\left(\frac{D}{H}\right) \leqslant 0.005\left(\frac{D}{H}\right)^2. \qquad (5.15)$$

This means that in the particular case where $D/H = 9$, an error amounting to about 0.4 or 4% can be allowed in D/H determination to obtain a 0.5% error in the deuterium concentration [448].

A method for analyzing a ternary hydrogen-deuterium-air mixture in an arrangement of the type shown in Fig. 85 is described by Broida and Morgan [449]. The air has no effect on the relative intensities of the hydrogen and deuterium lines. The $D/H+D$ ratio remains constant to within 0.07% upon addition of 4.8% air, and constant to within 0.1% if the amount of added air is 50%. Figure 86 shows the calibration curves of Broida and Morgan for determination of air in a hydrogen-air-deuterium mixture. The accuracy of analysis with the aid of these curves is about 0.1%, with the limit of detection for air being 0.01%. The curves were plotted for air concentrations below 6%, and deuterium concentrations above 90%, but the range can be extended considerably for both components.

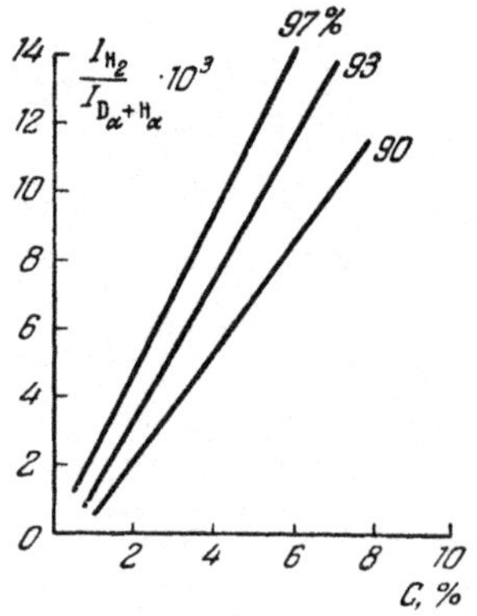

FIG. 86. Calibration curves for determination of air in an air-hydrogen-deuterium mixture, at varying deuterium concentrations.

It was found that the ratio of line intensities is not the same as the ratio of concentrations of hydrogen and deuterium in the mixture. The analysis was therefore based on calibration curves plotted for specified discharge conditions.

Zaydel' et al. [453] investigated the possible causes of discrepancies between the line intensity and isotope concentration ratios. They tested the effects of self-absorption, degree of dissociation, separation of mixture components, light scattering, overlapping of component lines, etc. They concluded that isotope separation in the capillaries was the major factor involved. By selecting the right pressure and flow rates, they were able to obtain line intensity ratios which were equal to ratios of isotope concentrations. This enabled them to develop a procedure for hydrogen determination in deuterium employing no standards. This procedure is valid within the concentration range of 2 to 90%. The apparatus and analytical results are described in [209, 290, 453].

Borgest and Zaydel' [209, 455] suggested the use of an interference polarization filter for determining traces of deuterium in hydrogen. The filter attenuated the stronger H_α line so that the edge of the line no longer interfered with determination of deuterium concentrations of the order of 0.01%. This analysis uses calibration curves.

Ostrovskaya and Zaydel' [454] proposed reducing the pressure in the discharge tube to 1-2 mm Hg in order to improve the resolution of hydrogen and deuterium lines. This led to a narrowing of the line shapes for both gases, since these shapes resulted from collision broadening. According to these authors, analysis without standards can be employed at concentrations down to 0.1%. The determination of lower concentrations requires standards. For deuterium concentrations close to the natural content in hydrogen

(0.015%) the error reaches 25%, but it decreases at higher concentrations. A visual method based on the use of a polarization deuterometer was developed for determining large deuterium concentrations in hydrogen.

Oganov and Striganov [451, 452] used standards to analyze ternary mixtures of hydrogen, deuterium and tritium.

The analysis of isotopic hydrogen-deuterium mixtures is also used to determine hydrogen in metals via the method of isotopic equilibration [290].

Determining the isotopic composition of helium

Such a method was developed by Artaud, Blaise and Gerstenkorn [457]. The isotope shift of the He λ 6678 Å line amounted to 0.5 Å. Therefore, high dispersion equipment was necessary (Fig. 87). Special techniques were used to obtain narrow spectral lines. The

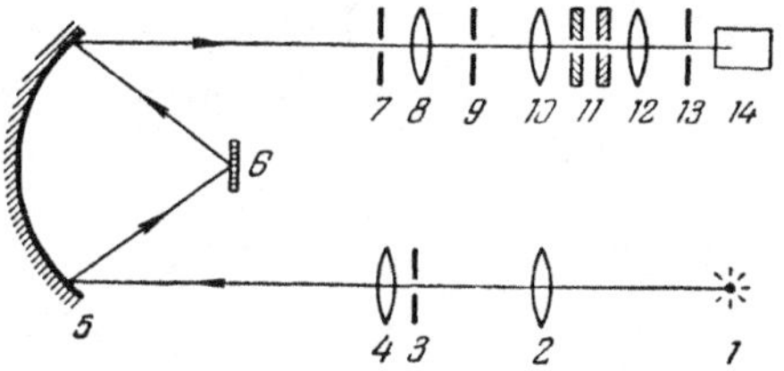

FIG. 87. Optical system of the arrangement for isotopic analysis of He, Li, Pb and U. 1—Light source; 2, 4—illuminating lenses; 3, 7—monochromator slits; 5—mirror; 6—diffraction grating; 8, 10, 12—projecting lenses; 11—Fabry-Perot interferometer; 9, 13—diaphragms; 14—photomultiplier.

glow was studied inside a hollow cathode cooled with liquid air. The discharge tube current uses a few milliamperes. The analysis was conducted without the use of standards. The ratio of concentrations of the helium isotopes was assumed equal to the ratio of areas under the spectral peaks. Special tests demonstrated that the instrument does not distort the intensity distribution, and that the line contours are subject to the Doppler effect. The Doppler broadening is inversely proportional to the square root of the atomic mass, and hence the ratio of concentrations of the helium isotopes is given by the expression

$$\frac{C_{He^3}}{C_{He^4}} = \frac{h_3}{h_4}\sqrt{\frac{\mu_{He^4}}{\mu_{He^3}}} = 1.15\frac{h_3}{h_4}, \tag{5.16}$$

where h_3 and h_4 are the peak heights in the recordings and μ_{He^4} and μ_{He^3} are the masses of the helium isotopes.

Determining the isotopic composition of carbon

This analysis is based on molecular spectra, in which the isotope shifts may amount to several angstroms. Broida and Fergusson [458] converted carbon to acetylene and excited the glow in an acetylene flame. Acetylene analysis was based on heads of bands $C^{12}C^{12}$ (λ 4737 Å) and $C^{12}C^{13}$ (λ 4744 Å). The spectra were detected by the photoelectric method, and calibration curves were plotted on the basis of standard mixtures. The analytical results were distorted by the presence of carbon dioxide in the air. In addition to carbon-containing samples, the isotopic composition of acetylene can be analyzed by this method.

An alternative method was developed by Zaydel' and Ostrovskaya [209, 459], using the arrangement for analyzing mixtures of hydrogen isotopes [453]. Carbon analysis was based on the CO λ 4124 Å band excited in a methane-oxygen mixture, and calibration curves based on the standards were used. It was necessary to take into account the overlap of the band $N_2\lambda$ 4142 Å. The method is suitable for working with concentrations ranging from 1 to 60%. The analytical accuracy is 5-7% at low concentrations (1-5%), and improves to 2-3% in the 10-60% concentration range. The amount of gas consumed in a single determination varies from 0.1 to 0.2 cm³. Analysis time is 10-15 min.

Determining the isotopic composition of nitrogen

The early determinations of isotopic composition of nitrogen were qualitative, and were designed to assist in regulation of isotope separation. Hoch and Weisser [460] were the first to develop a quantitative analysis. They compared the heads of molecular bands whose intensities were determined from photographs obtained at different exposures. They also determined the Schwarzschild constant. The analytical pairs used were λ 3159 and λ 3577 Å for $N^{14}N^{14}$, λ 3162 and λ 3573 Å for $N^{14}N^{15}$. An h-f oscillator provided the source of light. The concentration range was 0.38-99%, and the accuracy was about 2%. An analogous procedure was proposed by Zahn [463]. The mixture was analyzed in a high frequency discharge excited by an oscillator generating a frequency of 7 Mc. The discharge tube capillary was 1 mm I.D., and the gas pressure in the tube was about 2 mm Hg.

A photographic method for analyzing mixtures of nitrogen isotopes was developed by Gorbunov and Zagorets [461]. Over the concentration range of 0.38 to 8.5%, their analysis used an r-f oscillator and a quartz discharge tube with external electrodes. The nitrogen pressure was varied from 0.05 to 2.5 mm Hg. The results were accurate within 2-3%. A photoelectric method for determining nitrogen isotopes was used in [462, 464]. In [462] an r-f oscillator generating 2450 Mc was used, with the pressure maintained at 1 mm Hg. The detection was achieved by means of a monochromator with a resolution of 0.1 Å, and scanning rate of 5 Å/min. The analysis was either of the flow type or proceeded in a sealed tube. While the accuracy was higher in the first case, the quantity of gas required was also considerably greater (20 cm^3 at p = 1.5 mm Hg). The analysis time was 10 min. The analysis was based on the lines of the rotational structure of the 1.0 and 0.1 bands in the second positive system. Within the

0.38-90% concentration range, the analysis required calibration curves.

Zaydel' and Ostrovskaya [464] described a method for nitrogen isotopes useful over the 0.38-15% range, and employing no standards. The spectra were excited in a high frequency discharge at a frequency of 6 Mc. The heads of band 2.0 of the second positive system were used as the basis of analysis. An accuracy of about 5% was attained at concentrations close to the natural one (0.038%).

This completes the list of gaseous mixtures that can now be analyzed by spectroscopic methods. The range will doubtless be extended. The composition of any isotopic mixture can probably be determined by spectroscopic means. It is interesting to note that deuterium, the heavy isotope of hydrogen, was discovered by studying hydrogen spectra [465], and many other isotopes were discovered in the same manner.

Chapter VI

SPECTRUM ABSORPTION METHODS IN ANALYSIS OF GAS MIXTURES

30. GENERAL DESCRIPTION OF SPECTRUM ABSORPTION METHODS

The absorption methods for analysis of gas mixtures can successfully compete with procedures based on emission spectra. It is a matter of common knowledge that in recent years much attention has been focused on developing absorption methods for analysis of metals [374, 466-468]. A general description of such methods is offered by Walsh [374], who also discusses their merits as compared with other techniques. Walsh argued that an analysis based on absorption spectra can be considered as an absolute method, whereby the need for comparison standards is eliminated. An added and essential advantage is the elimination of the third component effect. Finally, the analytical sensitivity no longer depends on the excitation potential of the particular line. Thus, the sensitivity for sodium is the same as for zinc [466].

Walsh's argument that absorption sprectroscopy is an absolute method has, however, proved somewhat optimistic. For one thing the transition probabilities are known for some, but by no means all, resonance lines. Furthermore, absorption lines are in most cases distorted by the isotopic and hyperfine structure. This leads to errors in determining absorption coefficients and, therefore, in evaluating the absolute concentrations. It is necessary, moreover, to know the vapor pressure equation for the metals in the sample. Absorption spectroscopy of metals, especially for those melting at high temperatures, involves considerable experimental difficulties, since the absorption becomes appreciable only at high temperatures. While high temperature furnaces need not be used when working with gases, which simplifies the experimental technique, there are some specific difficulties which must be overcome. Thus, in analyzing inert gases whose absorption lines lie in the vacuum region, the entire spectrograph must be filled with the mixture, or else one requires cuvettes with windows transparent to vacuum UV. The experimental difficulties involved appear to have been holding back the development of methods for gas analysis in the vacuum spectral region. Absorption spectra for wavelengths of 1100 Å and higher can be obtained using cuvettes with lithium fluoride windows. These cuvettes are, however, useless for absorption spectroscopy of helium, neon and argon, since the wavelengths of the resonance lines of these gases are λ 584, 736 and 1048 Å. However, the problem should not be considered hopeless since manufacturing techniques are currently being developed for the preparation of solid transparent films designed for working in the far ultraviolet region of the spectrum.

Analysis of gases absorbing in the near UV, visual and IR regions is far less complicated. The most widely accepted methods use IR absorption spectra, which arise predominantly from rotational and rotational-vibrational molecular transitions.

Rotational spectra lie in the far infrared with wavelengths from a few hundred microns to several centimeters. This wavelength range is practically inaccessible to optical spectroscopy, and therefore radiofrequency methods must be employed. Analytical applications of microwave spectroscopy were reviewed by Zeil [469].

The rotational-vibrational spectra correspond to the wavelength range of 2 to 100 microns. They partially overlap the spectral region which can be detected without difficulty both on photoplates and with the aid of thermocouples and other devices.

Vibrational bands appear only in the spectra of molecules possessing a dipole moment. It follows that rotational-vibrational spectra cannot be obtained for symmetrical diatomic molecules; hence, the analysis in this case must be based on electronic spectra. The electronic molecular spectra correspond to quanta of considerably higher frequency, and can therefore be observed in the visible and UV spectral regions.

The fundamentals of analysis based on absorption spectra have been elaborated upon in a number of papers [317, 470, 471].

In view of the difficulties involved in analyzing the absorption of resonance lines of inert gases, a procedure based on absorption in excited gases was developed [433].

31. ABSORPTION IN EXCITED GASES AS A BASIS OF ANALYSIS

To determine impurities in inert gases, light is transmitted through an absorption discharge tube filled with an excited inert gas from a source discharge tube filled with the same gas. One thus obtains an attenuation or absorption of the light at several spectral line frequencies. According to [64, 433, 472], the maximum absorption occurs in lines which appear as a result of transitions from the upper to the metastable levels $^3P_0$ and $^3P_2$, as well as to the excited level $^3P_1$ (see the energy level diagrams in Section 1). The excited atom concentration depends strongly on the purity of the gas in the discharge tube. The presence of such impurities as nitrogen, hydrogen, oxygen, or water vapor reduces the concentration of excited atoms, due both to collisions of the second kind between the impurity atoms and the metastable atoms of the primary gas, and to the electron temperature drop. The decreased excited atom concentration, in turn, causes a decrease in the amount of absorption. This occurs only in the case where the first excitation potential of the impurity is lower than the excitation energy of the metastable levels of the primary gas.

To characterize the absorption process we shall introduce the *absorbance* A expressing the ratio of radiation absorbed by the sample to that incident upon the sample

$$A = \frac{I_S + I_a - I_0}{I_S} 100\%; \tag{6.1}$$

where I_S, I_a and I_0 denote, respectively, radiation emitted by the source discharge tube alone, the absorption discharge tube alone and the total radiation of both tubes operating simultaneously.

Absorbances of different neon lines at varying impurity concentrations are shown in Table 3. The variability of the absorbance of pure neon can be traced to differing conditions (different pressures and lengths of the discharge tube) under which measurements were carried out.

The dependence of absorbance on the impurity concentration provides a basis for plotting calibration curves for the determination of a readily excited impurity in inert gases, as well as for determining traces of an inert gas with a lower ionization potential in an inert gas having a higher ionization potential.

It will be remembered that what is determined is not the absorption of radiation by the impurities, but a diminution of the absorption by the neon atoms themselves. This is both the superiority and the weakness of the method. On the one hand, it fails

Table 3

Variation of absorbance with impurity concentration for various neon lines [433, 473]

Impurity	C, %	A (in %) for lines (in Å)			
		$^3P_2$		$^3P_0$	
		6143	6402	6266	6163
Nitrogen	0.00	60.7	70	43.4	40.7
	0.075	51,3	63	39.4	33.3
	0.15	49	63	34.1	27.7
	0.26	45	58	24.1	20.5
	0.39	21.8	27	9.8	7.9
Argon	0,00	100	100	100	90
	0.50	90	85	55	34
	1.0	76	80	35	25
	2.0	45	65	4	9
Hydrogen	0.00	48.3	61	31.4	26.3
	0.06	33.5	48	18.5	15.4
	0.12	12.2	17	4.5	6.0
	0.29	8.7	10	1.8	2.6
	0.17	11.2	14	3.7	4.0

to identify the individual impurities; on the other hand, this considerably simplifies the analytical procedure, since the absorption of neon lines can be measured more easily than that of the lines of the impurities.

The fact that lines of a fairly narrow spectral region (Neλ 6143—6506 Å; Arλ 6365—8014 Å) produce considerable absorption permits dispensing with spectroscopes and use of simple light filters. In addition, a procedure based on total radiation absorption is most adaptable to a simple system of photoelectric detection.

A schematic diagram of an arrangement for absorption analysis of gas mixtures [433] is shown in Fig. 88. The source discharge tube *1* is filled with the principal component of the mixture and gives off a very intensive radiation flux (the internal diameter of the capillary is several millimeters). The absorption discharge tube *3* is filled with the sample to be analyzed. The tube dimensions can be varied depending on the composition of the sample, since its purpose is to provide an absorption column of sufficient length. The source and absorption tubes are connected to sources providing, respectively, 30 mA and 10 mA, 1500 V DC current. An objective lens *2* directs a parallel light beam from the source through the absorption tube and light filter *4*; this beam is then focused (by objective *5*) on the photocathode of photomultiplier *6*. After passing through a DC amplifier, the photocurrent is indicated by galvanometer *7*. Either interference filters or a set of tinted glasses can be used as light filters. Interference filters, which operate in a relatively narrow passband (100–150 Å), give more accurate measurements than conventional colored glass filters. Colored glass filters invariably pass through a portion of the source radiation which is outside the absorption region. The result is that the accuracy of measurements is reduced. If in the above arrangement the absorption tube is filled with a gas mixture of known composition, one can plot a calibration curve. The concentrations of the gas to be determined are plotted on the abscissa, and the absorbance values along the ordinate. If then one measures absorbance, the concentration of the minor component (impurity) can be determined from the calibration curve.

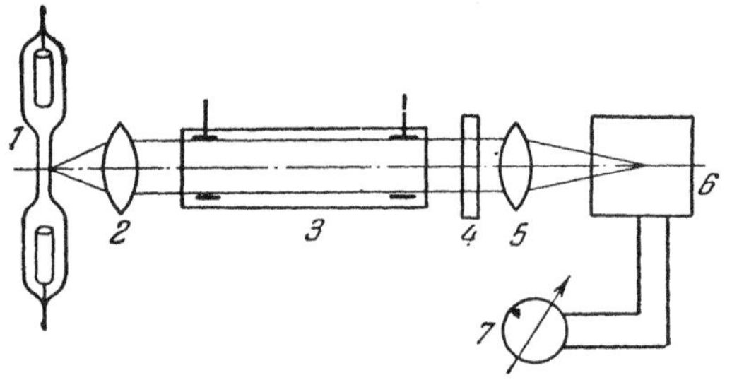

FIG. 88. Experimental arrangement for absorption analysis of gas mixtures.

The analytical procedure can be considerably speeded by calibrating the galvanometer readings directly in percent of the minor component.

Figure 89 shows the schematic diagram of a double beam unit (the notation is the same as in Fig. 88). The light from the source discharge tube (two parallel beams) passes through two identical absorption tubes *3* and *3′* and is detected by two independent photomultipliers *6* and *6′*. The luminous flux difference is indicated by the galvanometer. If the two absorption tubes and the source discharge tube are filled with the same (principal) component of the sample, the two fluxes will be equal and the galvanometer will indicate zero. If one of the absorption tubes is filled with the sample, the light flux passing through it will be less attenuated (the concentration of absorbing atoms is lower because of the presence of the minor component). The galvanometer will then indicate a flux intensity difference which is a function of the concentration of the minor component.

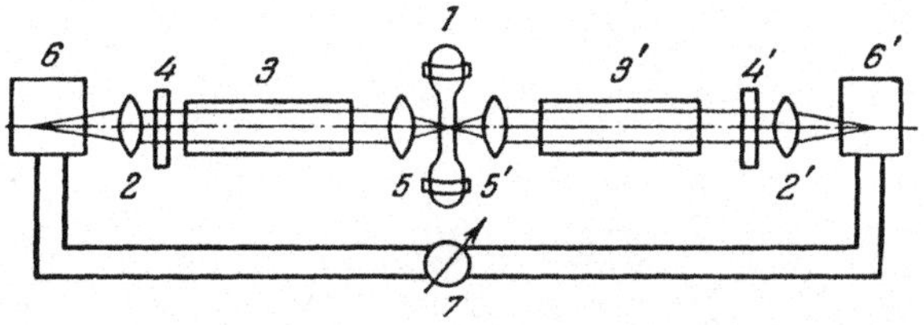

FIG. 89. Double beam unit for absorption spectroscopy of gas mixtures.

This procedure is most suitable for analyzing binary gas mixtures. It can be used in determining impurities in inert gases, when the ionization potentials of the impurities are lower than those of the primary gas (see Fig. 90). It can also be used for molecular gases (nitrogen, hydrogen, oxygen, carbon dioxide) whose ionization potentials are lower than the energies of the metastable levels of inert gases (see Fig. 91). Figures 90 and 91 were obtained at the

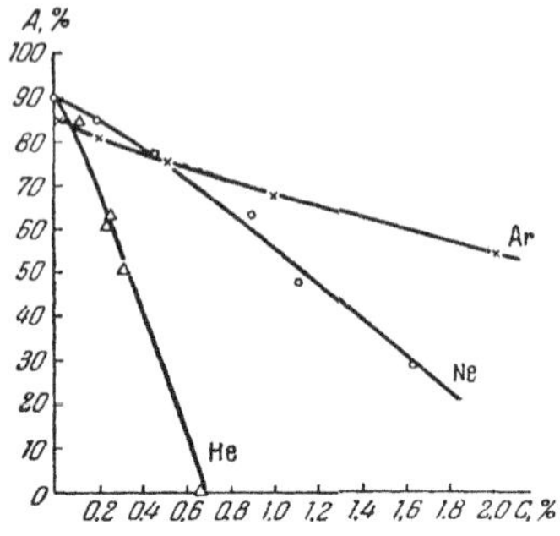

FIG. 90. Calibration curves for determination of nitrogen in inert gases.

following conditions: discharge tube 60 mm long, 2 mm I.D., pressure 6 mm Hg; 30 mA current; absorption discharge tube, 500 mm long, 20 mm I.D.; variable pressure, 1 to 5 mm Hg, depending on the specific analytical problem; 10 mA current. In plotting the calibration curves, special care was given to the selection of the spectral region and excitation conditions in the absorption discharge tube. The spectral region was usually one which corresponds to a maximum absorption. Since such regions were fairly wide, further restrictions were imposed by the available filters, as well as by the spectral sensitivity of the photo tube and photomultiplier cathodes.

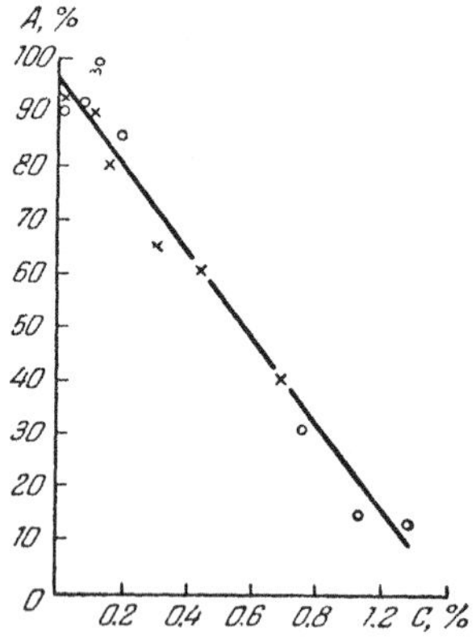

FIG. 91. Calibration curve for determination of argon in neon.

The curves for argon and neon in Figs. 90 and 91 were obtained with the aid of interference filters. In the case of argon the filter transmission peak was at λ_m7850 Å, and with neon the peak was at λ_m6300 Å. For helium, various combinations of glass filters were used. The passband thus obtained limited the spectrum to the infrared region beginning with λ8000 Å.

The sensitivity of analysis to changes in concentration increases markedly with the diameter of the absorption discharge tube and the pressure in it (see Fig. 92). It follows that impurities of the order of 0.001% to 0.01% must be determined at high pressures, using large-diameter absorption discharge tubes. In such cases the absorbance in a pure gas is rather low, and as a result the accuracy of the measurements is considerably reduced. To improve the accuracy, it is necessary to increase the length of the column of absorbing gas, i.e., the length of the tube.

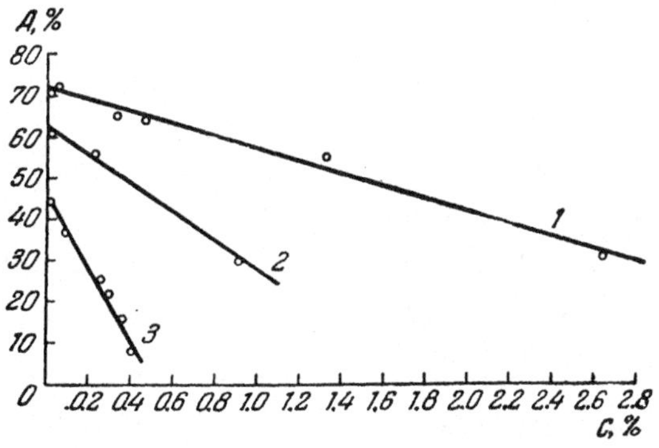

FIG. 92. Effect of pressure on the sensitivity of the analysis for nitrogen in argon. 1—p = 0.5 mm Hg; 2—p = 1.4 mm Hg; 3—p = 2.8 mm Hg.

Analysis of multicomponent mixtures based on measuring absorbance is possible only when one of the components of the mixture possesses a higher ionization potential than that of the principal component. If this is not the case, it becomes impossible to determine which component is responsible for the decrease in absorption.

The above procedure is a simple and fast test for the presence of impurities in inert gases, even though the percentage of the impurity may remain unknown.

Absorption of radiation in an excited gas was used as a basis for an analysis of a mixtured hydrogen isotopes [209, 446]. The source tube was filled with the sample, while the absorption discharge tube was filled with light hydrogen. Both tubes were connected to a source of DC current. The light in the source tube was modulated, and a special amplifier tuned to the modulator frequency was provided in the system so that only the emitted radiation was recorded. A marked absorption of the Balmer series was noted. The procedure can also be modified. Thus, the absorption, rather than the source, tube can be filled with the sample. Mixtures of isotopes of inert gases can also be analyzed by observing absorption in excited gases.

32. INFRARED ABSORPTION ANALYSIS OF GASEOUS MIXTURES

Infrared absorption spectra have been used in gas mixture analysis for more than 20 years and this application has been extensively reviewed [471, 474-476].* The analysis can be carried out with or without dispersion of the radiation.

Analysis involving the dispersion of radiated light is based on the fact that each constituent of the mixture has specific absorption bands. It is assumed that the other constituents do not absorb in this region. The concentration of the constituent in the sample is determined from the intensities of the absorption bands. Dispersion of the light is achieved by means of a spectrograph or a monochromator. Bolometers, thermopiles or photographic plates are used for detection and/or recording. The type and operating conditions of the radiation receiver materially affect the accuracy of the measurement [478]. This type of analysis is based on the application of Beer's law to each component of the mixture. However, Beer's law frequently does not hold exactly. Various

*For information on techniques of infrared spectroscopy, see general manuals and reviews [317, 477].

methods are available for correcting these deviations from the ideal behavior [479, 480].

Spectrographs or monochromators are used in cases where a high analytical accuracy is required. Such devices also require highly sensitive detectors.

A gas analyzer operating on the principle of a double beam spectrophotometer was proposed by Pemsler and Planet [481]. The spectrophotometer consists of two chambers, each of which is filled with the sample component which is being determined. One of the chambers is then closed off and the other is evacuated. Next, the sample is slowly introduced into the evacuated chamber until absorption is the same in both chambers, and the instrument needle is at zero. The pressure ratio in the two chambers then indicates the percentage of the desired constituent. The method is accurate to within 0.1-2%. An automatic prism-diffraction grating IR gas analyzer is also described in [482].

Analysis without dispersion of radiation is known as the integral absorption method [471]. Selectivity is attained either with the aid of a selective detector, a selective filter, or a selective emitter (source). The last involves the greatest design difficulties, but there is a possibility that the problem will be solved through the use of lasers. As a rule, a selective detector is the gas whose content is being determined; it is heated by the absorption of radiation, and the change in heat content is then indicated by a bolometer. The same gas, when giving off the absorbed radiation, can also act as a selective filter. This function can also be performed by an interference filter which separates a narrow spectral region. Interference filters are widely used in conventional gas analyzers [483, 484].

The Parsons-Irland-Bryan gas analyzer [484], designed for hydrocarbon determination in exhaust gases, is shown in Fig. 93.

The unit consists of a ribbon filament tube *1*, a quartz window *2*, an aluminum tube *3* for passage of the gas, a quartz lens *4*, filters *5* and a thermocouple *6*. An interference filter with a half-thickness of 0.15 micron separates out the 3.43 micron absorption band, while a second filter chops off the wavelength range below 2.4 microns. To determine *n*-hexane, an analytical curve was plotted (with an accuracy of 0.1%) for the concentration range of 0.1 to 0.2 mole%. The thermocouple readings, which indicate the fraction of transmitted radiation, vary with concentration.

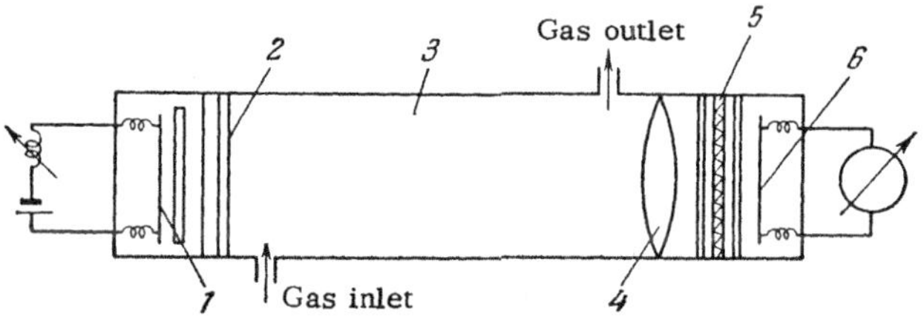

FIG. 93. Schematic diagram of the Parsons-Irland-Bryan gas analyzer.

Interference filters which isolate individual lines of the rotational band structure are also used in the Harrison and Reid [483] unit (see Fig. 94) designed for hydrogen fluoride (partial pressure range of 0.02-20 mm Hg) determination in corrosive gas mixtures at a total pressure of 1 atm. The principal constituent of the mixture is nitrogen, and uranium hexafluoride (UF_6) is the minor component. The apparatus consists of a tungsten lamp connected to a voltage stabilizer, a system of lenses, a shutter and a detector. Two beams are transmitted through the device, where one undergoes absorption while the other does not; both beams then impinge on the detector. Two filters are provided—a fixed wide-band one (with 2-2.5 micron passband) and a narrow-band one, with a half-thickness of 0.004 micron. The last filter is mounted on a modulator vibrator oscillating at 10 cps. A narrow spectral region is scanned; this is achieved because of the unique capacity of an interference filter to change its output λ_m with a change in the angle of incidence. When the vibrator is disconnected, the filter transmits line λ_0 (one of the lines of the rotational structure of HF); when the filter is in circuit, the line transmitted is λ_1 (it lies between the rotational structure lines). The differential reading for these two wavelengths indicates the partial pressure of the hydrogen fluoride.

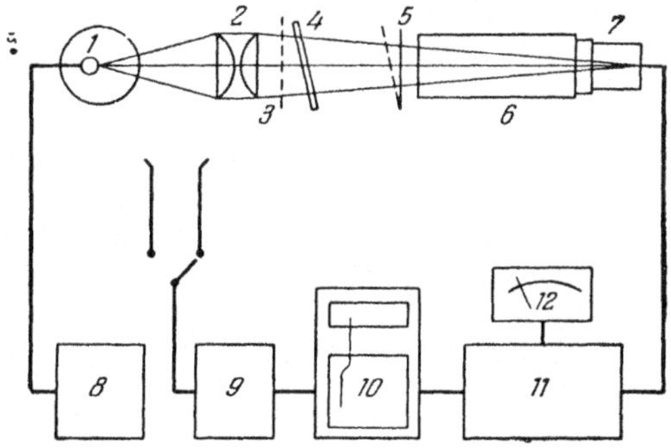

FIG. 94. Gas analyzer with mobile filter. 1—Light source; 2—lenses; 3—shutter; 4—filter; 5—oscillating filter; 6—chamber with the sample; 7—detector; 8—light source power supply; 9—oscillator; 10—recorder; 11—amplifier; 12—measuring instrument.

A gas analyzer with selective filter (see Fig. 95), where the filter is the analyzed gas itself, was proposed by Pfund [485]. As seen from the diagram, the two beams emerging from the light source *1* pass through sample *2*, then through a gas filter whose excitation bands coincide with those of the gaseous component analyzed, and finally through chambers *3* and *4* and onto the two detectors *5* and *6* connected in opposition to each other; a measuring device *7* is inserted in the circuit. Chamber *3* is filled with the gas which is being determined; the object is to absorb completely the radiation at frequencies characteristic of this gas and prevent it from reaching detector *5*. Chamber *4* is filled with any gas which does not absorb at these frequencies. Detector *5* thus indicates total radiation, except the fraction emitted by the gas to be determined. Detector *6*, on the other hand, indicates the total radiant flux. The difference in intensity between the two signals is thus a function of the unknown concentration of the gas. The higher the unknown concentration, the closer will be the readings of detectors *5* and *6* (since the absorption in chamber *3* will tend to decrease—whatever radiation could be absorbed there is already absorbed by the sample).

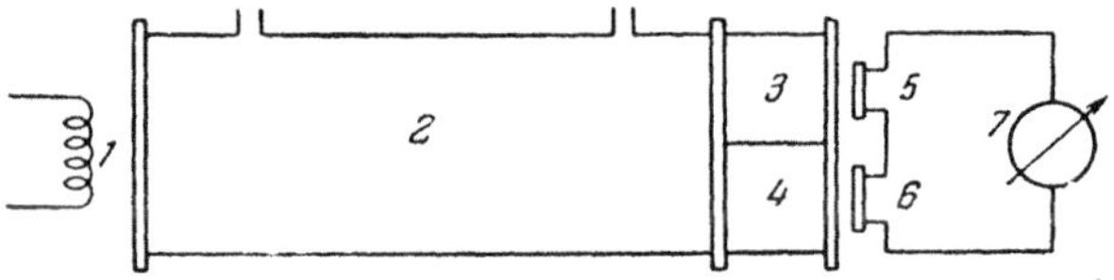

FIG. 95. The Pfund gas analyzer.

The chief merit of this analyzer is its simplicity. Its disadvantage is that the signal (relative to total radiation) is small. Consequently, high sensitivity detectors and stabilized light sources must be used.

The most widely used units employ selective detectors. This type of analyzer, first proposed by Veyngerov [486], was subsequently modified (see Fig. 96) by Luft [487]. Light beams are radiated from spiral coils *1* and *2*, pass through a modulator

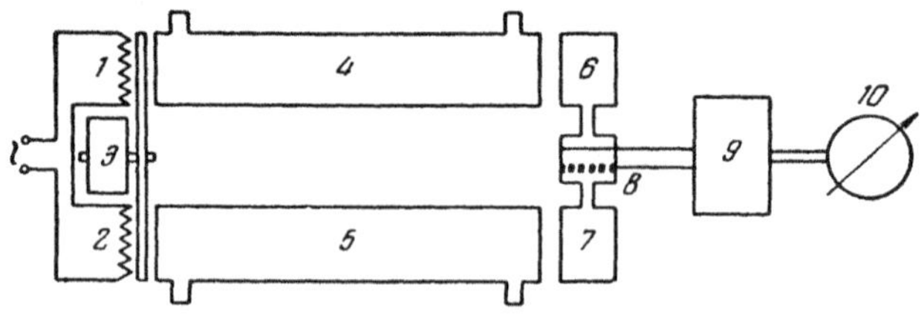

FIG. 96. The Luft gas analyzer.

stage *3*, and are then directed, respectively, through chambers *4* and *5*. The former is filled with the sample; the latter is empty and can be filled with any standard mixture.

Chambers *6* and *7*, separated by a membrane strain gage *8*, are filled with the gas whose concentration is unknown (selective detector). Because the absorption is not the same in chambers *4* and *5*, there exists an absorption differential between chambers *6* and *7*. If chamber *5* is empty, then because of previous absorption in *4* the absorption in *6* is smaller than in *7*. As a result, the gas temperature in the two chambers is different and a pressure differential is created. Unequal pressures acting on the strain gage can be converted to an alternating current, which is amplified in *9* and recorded by instrument *10*. The composition of gases in the sample can thus be continuously monitored.

Gas analyzers operating on the principle of selective detection are far more sensitive (but are also more complex) than analyzers using selective filters.

Another gas analyzer whose operation does not involve light dispersion was designed by Veyngerov [486] and used for analyzing molecular gases and vapors. This *optico-acoustic* procedure is based on the fact that a "sound" will be produced in the gas absorbing infrared radiation, if the radiant beam is interupped at an audio frequency rate. In the process of absorption, the gas becomes heated, and its pressure rises. Because the radiant flux is interrupted at an audio frequency rate, the resulting pressure ripple in the gas produces a sound, which is recorded by a microphone. The sound intensity depends on the capacity of a given gas for absorbing the radiation.

A block diagram of the optico-acoustic analyzer is shown in Fig. 97. The source of the radiation is a platinum strip *4* (about 0.05 mm thick, 20 mm long and 15 mm wide) enclosed in a sealed metal case *3*. The radiant flux emerging from the case through the uviol glass window *5* is interrupted with a frequency of 200 cps by means of a flicker shutter *1* mounted on the shaft of an electric motor *2*. On passing through a fluorite window *6* the radiation falls into the audio-optical chamber *8*. The gas sample is introduced into the same chamber and flows through channels *7* and *10*. The interior surfaces of the chamber are gold-plated and polished, hence losses due to reflection from the walls are negligible. The moving-coil microphone *9* is isolated from the chamber by a partition made of fine aluminum foil. The current set up in the microphone is amplified by a two tube DC amplifier *11*, and is indicated by galvanometer *12*.

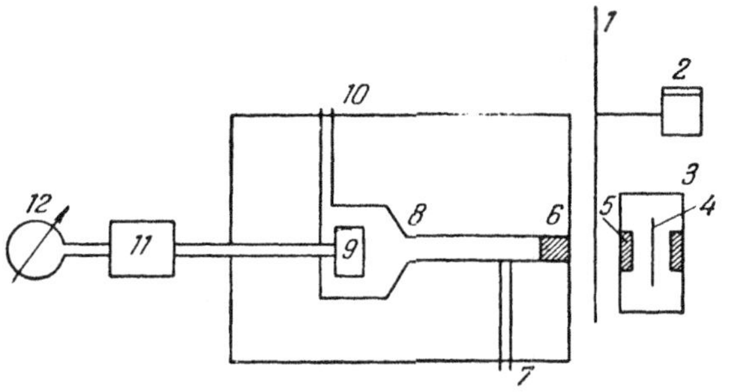

FIG. 97. Block diagram of an optico-acoustic gas analyzer.

The least complicated analysis by this method involves a binary mixture consisting of an absorbing and a non-absorbing gas. Galvanometer readings will in this case be a direct function of the concentration of the absorbing gas. The calibration curve is plotted with mixtures of known concentrations. Thus, calibration curves for carbon dioxide in air are shown in Fig. 98. The air is considered the non-absorbing component, since neither nitrogen nor oxygen absorbs IR radiation (the method is therefore inapplicable to analysis of oxygen-nitrogen mixtures, nor can it be used for hydrogen, since the hydrogen molecules do not absorb in the near infrared). The calibration is steep at low concentrations, then the slope becomes flatter, and finally, at high concentrations, α (the galvanometer deflection) becomes virtually independent of the CO_2 concentration. The reason for this is that the energy absorption is determined largely by the mid-region rather than the fringes of the band. At high concentrations, nearly all of the mid-band is absorbed, so that increased absorption of the fringes contributes very little additional energy.

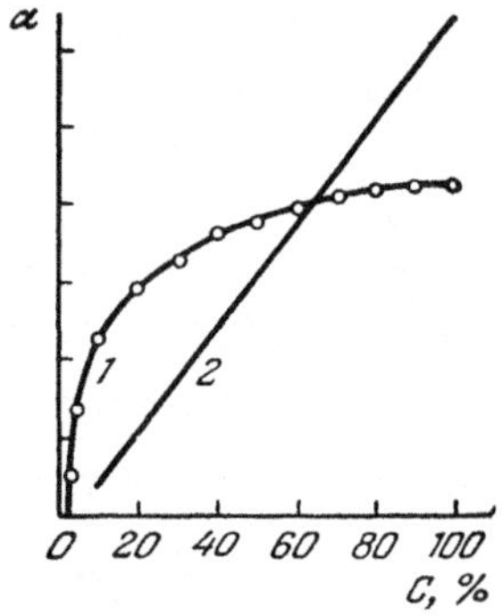

FIG. 98. Calibration curve for determining carbon dioxide - air mixtures. 1—without filter; 2—with filter.

Two methods were proposed by Veyngerov and Luchin [488, 489] for improving the analytical sensitivity of this method at high concentrations. In the first procedure, a filter is inserted between the absorption chamber and the light source. The filter—a tube with reflecting walls, provided with windows transparent to radiation—is filled either with the gas whose concentration in air is to be determined, or with a mixture containing a known amount of this gas. The fraction of radiation most intensively absorbed by the gas is blocked by the filter. As a result, the light flux entering the measuring chamber will be absorbed to a far lesser extent, and the system will once again show its initial rectilinear characteristic (curve *2* in Fig. 98).

The second method is based on reducing the length of the absorbing gas column. If this gas column is long, then the extent of radiant flux absorption is already high at low concentrations, so that an increase in concentration will not appreciably affect the amount of heat transmitted to the gas. If, on the other hand, the column is short, the total heat transmitted to the gas will vary linearly with the concentration, and this relationship can be taken as a basis for an analytical curve.

The filter method is suitable for analyzing mixtures consisting of two gases capable of absorbing radiation. This is true also in the case when the respective absorption spectra are quite dissimilar. A filter inserted between the absorption chamber and the radiation source traps the light flux in the spectral region corresponding to the absorption range of one component; then the sound will be produced by the other component of the mixture. The method is not always practicable in analyzing multicomponent mixtures.

A somewhat modified model was proposed by Veyngerov and Luchin [488, 489] for the analysis of multicomponent gas mixtures. The sample ($A + B + C$) is passed through an absorption tube situated between the chamber and the radiation source, rather than through the optico-acoustic chamber. The latter chamber is filled with the mixture component to be determined, for instance, A. The radiation passing through the absorption tube will then be absorbed by the sample. The light flux falling into the optico-acoustic chamber is thus reduced, and as a result the sound intensity is attenuated. The attenuation increases with the concentration component A in the sample. The concentration of A is found from a calibration graph plotted for this component. A multicomponent mixture can thus be fully analyzed by filling the chamber successively with each component, and determining the concentrations with the aid of respective calibration curves. For convenience it is best to have as many individual chambers as there are components in the sample. In the measurement procedure, the chamber filled with the desired component is placed after the radiation source and the absorption tube. The needle of the output instrument will then indicate the concentration of this component in the sample. A monochromatic radiation source [470] must be used when analyzing samples containing components with overlapping absorption bands.

The principal advantages of Veyngerov's gas analyzer are its low limit of detection (10^{-3}%) and rapid response.

The theoretical errors inherent in the optico-acoustic gas analyzer were discussed by Sall' [490].

Infrared gas analyzers were also successfully used in determinations of carbon and nitrogen isotopes [491-493].

33. ABSORPTION ANALYSIS OF GAS MIXTURES IN THE ULTRAVIOLET

Until recently, UV absorption was rarely used for analysis of gases. The reason for this is that very few gases absorb in the near ultraviolet (2000-4000 Å), while the vacuum region, as mentioned earlier, was practically inaccessible to spectroscopic analysis.

An automatic, double beam UV gas analyzer (Fig. 99) has been designed by Glasser [494]. Its photomultipliers are hooked up in a bridge circuit. The analysis is of the flow type.

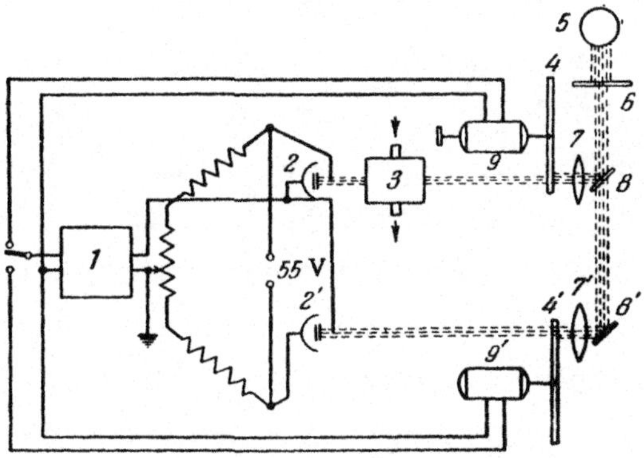

FIG. 99. Circuit diagram of Glasser's automatic gas analyzer. 1—Electrometer; 2,2'—radiation detectors; 3—absorption stage; 4,4'—modulators; 5—light source; 6—filter; 7,7'—condenser lenses; 8,8'—mirrors; 9,9'—motors.

The analyzer uses a filter or a monochromator, depending on the specific analytical problem. The filters employed are of standard type, and any wavelength from 2200 to 12,000 Å can be used. The length of the absorption cell may be varied from several hundrdths of a millimeter to 200 mm. This enables the analyst to work with both low and high concentrations. Therefore the unit is so sensitive that it can detect concentrations of the order of 10^{-4}%.

This analyzer has been used to determine hydrogen sulfide (λ 3130 Å), nitrogen dioxide (λ 3650 Å and λ 4050 Å), chlorine (λ 3340 Å and λ 3650 Å), as well as phosgene (λ 2540 Å and λ 2300 Å). The ozone concentration in air can be determined from ozone bands in the 2500-3000 Å region [495, 496], the limit of detection being $2 \cdot 10^{-6}$%. A round-the-clock automatic recorder was developed for this application.

Quantitative analysis based on absorption in the UV region may prove an important tool in studying the kinetics of chemical reactions, for example, the reactions of atomic gases [497].

Various analytical techniques have been developed for working in the vacuum region. Chubb and Friedman [498] determined the water vapor concentration in air on the basis of the absorption of the hydrogen line L_α (1215 Å) in the λ 1220 Å band (molecular nitrogen and molecular oxygen are almost transparent in this region). A photon counter sensitive within a very narrow region near 1216 Å served as the detector, eliminating the need for a monochromator. According to the authors, this method is suitable for measurement of traces of molecular oxygen in rare gases, nitrogen and hydrogen. Garton, Webb and Wildy [499] determined the water content of nitrogen, oxygen and carbon dioxide, on the basis of the L_α (1216 Å) absorption in the λ 1220 Å band. The hydrogen line L_α was excited in an electrodeless discharge at 20 Mc. The absorption vessel had lithium fluoride windows, and the radiation was detected by means of a tungsten-cathode photomultiplier sensitive in the range starting with 1400 Å. Two absorption tubes 1 cm and 42 cm long were used. Sensitivity was improved by means of aluminum mirrors, which caused the light to be repeatedly transmitted through the absorption cell. The limit of detection was 10^{-4}% for nitrogen, 10^{-3}% for oxygen and 10^{-2}% for CO_2.

Oxygen analysis based on absorption in the Schumman region was described by Prugger and Ulmer [500]. The oxygen was determined in the presence of nitrogen, carbon dioxide and water vapor, the oxygen content varying from 11 to 20%. Water vapor absorbs in the same spectral region as oxygen, which complicates the analysis. While this reduces the accuracy, the analysis is still feasible, since in one spectral region (1600–1350 Å) oxygen absorbs more intensively than does water vapor and, in addition, its concentration in the sample is several times higher. The method is thus suitable only for determining high concentrations of oxygen. The accuracy of analysis is about 1%. The emitter (source) and absorption element are so selected that there will be no need for a monochromator. A hydrogen lamp with a thin quartz window provides a continuous spectral source, while the tungsten cathode acts as a selective pickup. The arrangement is adaptable to a continuous analysis; besides this is necessary because of rapid changes in the oxygen content of samples.

From the examples cited above it is evident that the vacuum UV region can be successfully used for determining the composition of gaseous mixtures.

Spectroscopic determinations in the UV can also be carried out with the aid of optico-acoustic analyzers [501]. The optico-acoustic effect in the UV region has been observed in nitrogen, oxygen and acetylene.

APPENDIX I

Sensitive lines and stable heads of bands*

Wavelength, Å	Intensity	Excitation potential, V
	Argon	
8115.31	5000	13.1
7503.87	700	13.5
7067.22	400	13.3
6965.43	400	13.3
4348.00	500	19.5
	Hydrogen	
6562.79	2000	12.1
4861.33	500	12.7
4340.47	200	13.0
	Helium	
5875.62	1000	23.0
4471.48	100	23.7
3888.65	1000	23.0
	Oxygen	
7771.95	1000	10.70
6158.21	1000	12.75
5330.65	500	13.06
4368.30	1000	12.36
3947.33	300	12.28
	Krypton	
5870.92	3000	12.1
5570.29	2000	12.1

*Letters v and r following wavelength values tabulated for band heads indicate that the band tends to spill over toward the violet or the red portion of the spectrum. Line intensities are taken from tables in [6] and head band intensities from tables in [20].

Wavelength, Å	Intensity	Excitation potential, V
	Xenon	
4671.22	2000	10.9
4624.27	1000	11.0
4500.98	500	11.1
	Neon	
6402.25	2000	18.6
5852.49	2000	19.0
5400.56	2000	19.0
	Nitrogen	
8911.6 v *	10	
6544.8 v *	10	
6468.5 v *	10	7.3
6394.7 v *	9	
5755.2 v *	7	
4059.4 v **	8	
3998.4 v **	9	11
3804.9 v **	10	
	Carbon monoxide	
6079.9 v ***	8	
5198.2 v ***	10	10.8
4835.3 v ***	10	
4510.9 v ***	10	
4393.1 ****	8	10.4
2505.7 ****	7	
	CH	
4312.5 v	10	2.86
	Cyanogen	
6631.6 r	9	
6478.7 r	10	1.35
6332.2 r	9	
5858.2 r	9	
4197.2 v	9	
4181.0 v	8	
3871.4 v	7	
	Carbon	
5635.4 v	8	2.4
5540.7 v	6	
5165.2 v	10	

*Positive system I.
**Positive system II.
***Ångström bands.
****Positive system III.

APPENDIX II

Pairs of Analytical Lines Used in the Quantitative Analysis of Gas Mixtures

The table summarizes the basic data obtained by spectral analysis of gas mixtures. Concentration ranges in many instances can be widened. The listed conditions of mixture excitation were not always optimum. In many instances they were dictated by specific experimental conditions (e.g., the small quantity of available gas). Practical considerations must therefore be taken into account in working out an analytical procedure (see Section 16). Where no literature source is indicated, the method was developed at the Physics Research Institute of Leningrad State University.

Concentration range, %	Analytical line pair, Å	Conditions of sample excitation	Standard deviation, %
	Analysis of impurities in nitrogen		
	Determination of argon		
0.1-20	Ar II 4806-N II 4810 Ar II 4806-N II 4743 Ar II 4736-N II 4810 Ar II 4736-N II 4743	p = 0.2 mm Hg d = 0.4-0.5 mm i = 400 mA f = 6 Mc/s	6
0.1-1	Ar I 7504- N I 7469	p = 0.2 mm Hg d = 0.5 mm i = 400 mA f = 6 Mc/s	
0.5-20	Ar II 4610-N_2 4600	High frequency hollow cathode $p = 6 \cdot 10^{-2}$ mm Hg i = 300 mA	
Limit of detection 0.8%	Ar II 4806	Ring discharge p = 0.2 mm Hg d = 30-50 mm λ = 60 m V = 3000 V C = 0.01 μF	

Concentration range, %	Analytical line pair, Å	Conditions of sample excitation	Standard deviation, %
0.1-98	Ar II 4736-N_2^+ 4709 Ar II 4765-N_2^+ 4709	p = 0.4-2 mm Hg Hollow cathode i = 40-100 mA	
0.90-1.1 [361]	Ar I 4159-N_2 3998	p = 2-10 mm Hg f = 150 Mc/s	
Determination of hydrogen			
0.1-10	H_3 4861-N_2 4800	p = 0.5 mm Hg d = 0.5 mm i = 200 mA f = 6 Mc/s	15
Determination of helium			
0.04-1	He 5876-N II 5667	Pulse discharge p = 8 mm Hg d = 12 mm V = 3000 V C = 30 μF	7
0.3-2	He 5876-N II 5667	p = 0.4 mm Hg d = 0.3 mm i = 400 mA f = 6 Mc/s	10
0.1-1	He 5867-N_2 5854	p = 0.5 mm Hg d = 0.5 mm i = 200 mA f = 6 Mc/s	
1.5-12	He 5876-N_2 4815	p = 1.2 mm Hg d = 0.4 mm i = 400 mA	
Determination of neon			
0.1-30	Ne 6929-N I 7424	p = 0.4 mm Hg d = 0.3 mm i = 400 mA	
Determination of carbon dioxide			
0.5-30	CO 4835-N_2 4815 CO 4835-N II 4601	p = 0.3 mm Hg d = 5 mm i = 350 mA	6
	CN 3883-N II 4601 CN 4216-N II 4601	p = 0.4 mm Hg d = 5 mm i = 350 mA	27

Concentration range, %	Analytical line pair, Å	Conditions of sample excitation	Standard deviation, %
51-95 [425]	CO 3127-N 2159	d = 5 mm p = 2.5 mm Hg f = 11 Mc/s	1.5%
Analysis of impurities in argon			
Determination of nitrogen			
0.001-0.1	N_2 3998-Ar I 4251	p = 10-20 mm Hg d = 10 mm i = 200 mA	7.5
0.1-10	N_2 3998-Ar I 4251	p = 0.5-5 mm Hg d = 5 mm i = 200 mA	
Determination of helium			
0.05-1	He 5876-Ar II 5145	Pulse discharge p = 5 mm Hg d = 10 mm V = 3000 V C = 72 μF	
0.1-1	He 5876-Ar I 6416 He 5876-Ar II 5062	p = 0.9 mm Hg d = 2 mm	
Determination of oxygen			
0.05-1	O 7772-Ar I 7724	p = 0.5 mm Hg d = 1 mm f = 6 Mc/s	
Analysis of impurities in hydrogen			
Determination of nitrogen			
0.05-3	N_2 3755-H_2 3924 N_2 3805-H_2 3924	p = 1 mm Hg d = 0.5 mm i = 150 mA	4.5
Determination of helium			
0.1-1	He 4471-H_γ 4340	p = 0.5 mm Hg d = 0.3 mm i = 350 mA	15
Determination of oxygen			
0.05-3	O I 7772-H_2 7789*	p = 1 mm Hg d = 0.5 mm i = 150 mA	9

*The mixture was enriched by passing through a palladium capillary.

Concentration range, %	Analytical line pair, Å	Conditions of sample excitation	Standard deviation, %
	Analysis of impurities in helium		
	Determination of nitrogen		
0.04-0.4	N_2 3020-Ar I 4300*	Positive column of glow discharge p = 0.5-8 mm Hg d = 3 mm i = 25 mA	
0.002-0.01	N_2^+ 4600-He_2 4626 N_2^+ 4600-He_2 4648 N_2 3998-He I 3889	p = 600 mm Hg d = 1 mm i = 300 mA f = 6 Mc/s	
0.001-0.1	N_2^+ 4278-He I 5047	p = 18 mm Hg d = 1 mm f = 6 Mc/s	
	Determination of argon		
0.005-0.1	Ar 4400-He 4388 Ar II 4426-He 4388	p = 200 mm Hg d = 0.5 mm f = 6 Mc/s	
1.5-10	Ar I 4259-He 4144 Ar I 4259-He 4121 Ar I 4159-He 4144	p = 0.6 mm Hg d = 2 mm i = 300 mA	6
0.5-2 [294]	Ar I 4159-He 4471	Glow discharge Signal lamp SN-1 p = 30 mm Hg	
	Determination of hydrogen		
0.05-2	$H\alpha$ 6562-He 6678 $H\beta$ 4861-He 4922	p = 2 mm Hg d = 10 mm i = 300 mA f = 6 Mc/s	7.5
0.001-0.08 [381]	$H\beta$ 4861-Ar I 4300*	Positive column of glow discharge p = 0.5-8 mm Hg d = 3 mm i = 25 mA	
0.001-0.1 [384]	$H\beta$ 4861-He 4922 $H\beta$ 4861-He 5047	p = 18 mm d = 1 mm f = 6 Mc/s	30

* Argon was introduced as an internal standard.

Concentration range, %	Analytical line pair, Å	Conditions of sample excitation	Standard deviation, %
Determination of oxygen			
0.2-1.2 [381]	O I 3964-Ar I 4300*	Positive column of glow discharge p = 0.5-8 mm Hg d = 3 mm i = 25 mA	
Determination of methane			
0.002-0.1	CH 4312-He 4388	p = 10 mm Hg d = 8 mm i = 25 mA	12
Determination of neon			
0.002-0.2	Ne I 6402-He 6678	p = 10 mm Hg d = 2-10 mm i = 300 mA	7.5
0.2-2	Ne I 6402-He 6678 Ne I 6507-He 6678 Ne I 6383-He 6678	p = 6 mm Hg d = 2 mm i = 300 mA	7.5
2-16	Ne I 5341-He 5048 Ne I 5341-He 5876 Ne I 5401-He 5048 Ne I 5401-He 5876 Ne I 5038-Ne 5401	p = 6 mm Hg d = 2 mm i = 300 mA	6
18-75	Ne I 5401-He 4922 Ne I 5852-He 5876 Ne I 5038-He 4922	p = 0.8 mm Hg d = 2 mm i = 300 mA	4-6
Determination of carbon monoxide			
0.01-1 [381]	CO 2296-Ar I 4300*	Positive column of glow discharge p = 0.5-8 mm Hg d = 3 mm i = 25 mA	
Analysis of impurities in oxygen			
Determination of nitrogen			
0.1-10	N_2 3998-O I 3947	p = 0.5 mm Hg d = 5 mm i = 300 mA	4.5

* Argon was introduced as an internal standard.

Concentration range, %	Analytical line pair, Å	Conditions of sample excitation	Standard deviation, %
Determination of krypton			
0.1-10	Kr I 5871-O I 5331 Kr I 5871-O I 6158 Kr I 5570-O I 6158	p = 0.8 mm Hg d = 0.5 mm i = 300 mA	
Analysis of impurities in xenon			
Determination of krypton			
0.2-14	Kr I 5871-Xe I 5875 Kr I 5871-Xe I 5895 Kr I 5871-Xe II 6036 Kr I 5570-Xe II 5419 Kr I 5871-Xe II 5976	p = 0.3 mm Hg d = 0.4 mm i = 300 mA	15
Analysis of impurities in krypton			
Determination of argon			
0.3-1	Ar I 6965-Kr I 6904* Ar I 7067-Kr I 6904	p = 0.3 mm Hg d = 0.5 mm i = 300 mA	12
Analysis of impurities in neon			
Determination of argon			
1.5-10	Ar I 4510-Ne I 5038 Ar I 4510-Ne I 5080 Ar I 4510-Ar II 4736	p = 1.3 mm Hg d = 2 mm i = 350 mA	4.5
2-8 [294]	Ar I 4191-Ne I 5341	Glow discharge in pilot tube p = 30 mm Hg	
Determination of helium			
0.1-1	He 5876-Ne I 5882	p = 0.5 mm Hg d = 0.5 mm i = 300 mA	
3-15	He 4922-Ne I 5074 He 5016-Ne I 5038	p = 0.5 mm Hg d = 2 mm i = 300 mA	

* The krypton contained 0.1% xenon.

Concentration range, %	Analytical line pair, Å	Conditions of sample excitation	Standard deviation, %
Analysis of impurities in carbon dioxide			
Determination of nitrogen			
0.5-25 [425]	CN 3883-CO 4124 N_2 3998-CO 4124 N_2 3998-CO 4392	p = 0.5 mm Hg d = 5 mm i = 350 mA	4.5
4-40	N_2 3159-CO 3242	p = 2.5 mm Hg	5
Determination of water vapor			
2.5-80 [434]	OH 3064	Ultrahigh-frequency discharge in TR and ATR tubes	

APPENDIX III

Soviet spectroscopic equipment described in this book: Basic Characteristics [209]

Model	Spectral region, Å	Optical dispersing system	Relative aperture		Focal length, mm		Optical system*	Reciprocal of linear dispersion in Å/mm for λ (in Å)							
			of collimator	of camera	of collimator	of camera		2000	3000	4000	5000	6000	7000	8000	10,000
ISP-22 ISP-28	2000–7000	Cornu prism	1:15	1:25	600	850	Fig. I	3.9	13.5	31.5	58.0	110			
ISP-51	3600–10,000	Two 60° prisms and one Abbe prism	1:5 1:5 1:10	1:3 1:5 1:13	300 300 600	120 270 800	Fig. II			42 19 6.3	105 47 17.5	196 87 30	336 149 50	441 196 66	770 342 115
ISP-51A	3600–10,000	Two 60° prisms and one Abbe prism	1:23		1300		Fig. III			2	5	9.3	16	21	37
KS-55 KSA-1	2000–10,000	Interchangeable quartz and glass 30° prisms	1:40 1:25		quartz 1750 glass 1725		Fig. IV	1.2	4.6	11.5 5.5	quartz 21 glass 12	34 21.5	33	47	85
DFS-3 DFS-13	2000–10,000	Flat grating, 600 rulings/mm, N = 7200 or 1200 rulings/mm N = 144,000	1:40		4000		Fig. V	For a 1200 rulings/mm grating, 1st order 4–4.5							
UM-2	3800–10,000	Constant-deviation 60° prism	1:6		280		Fig. VI								
MS	4000–7000	Constant-deviation 60° prism	1:4.5		130										
DM	2000–10,000	Flat grating, 600 rulings/mm			600		Fig. VII								
SL-3	4000–7000	Three 60° prisms	1:8		190	380	Fig. VIII								

* See following pages.

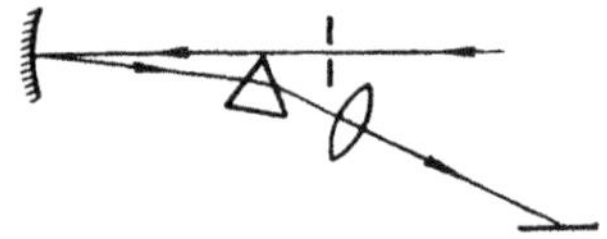

Fig. I

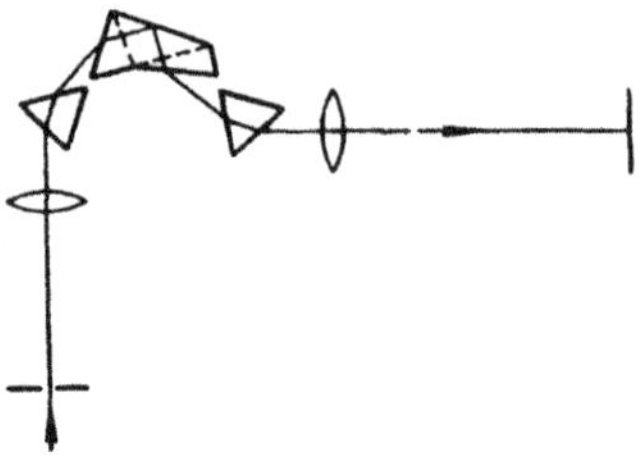

Fig. II

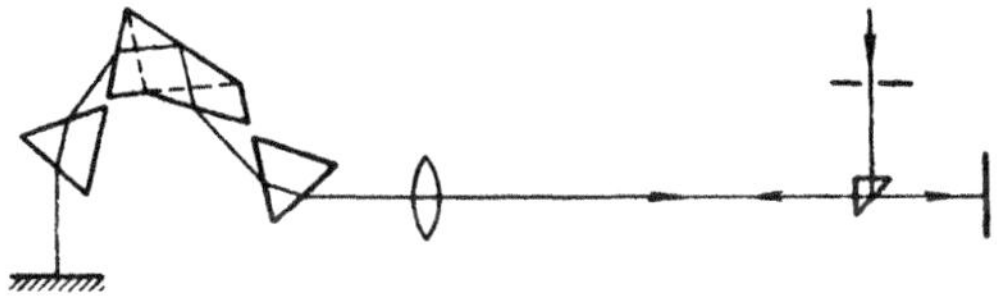

Fig. III

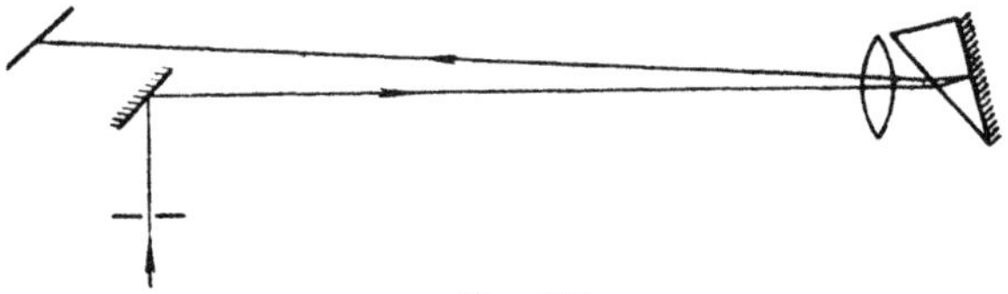

Fig. IV

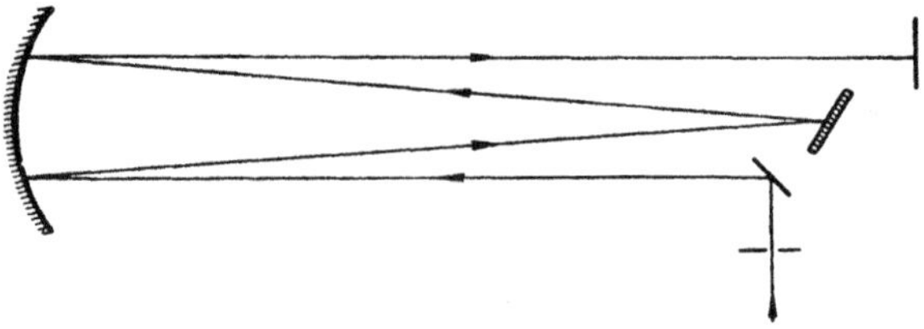

Fig. V

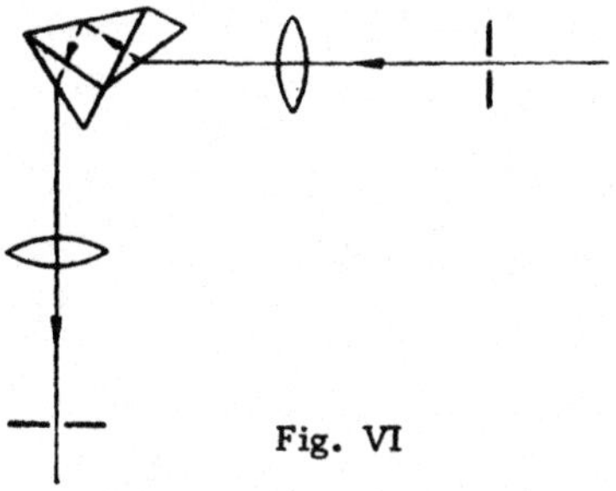

Fig. VI

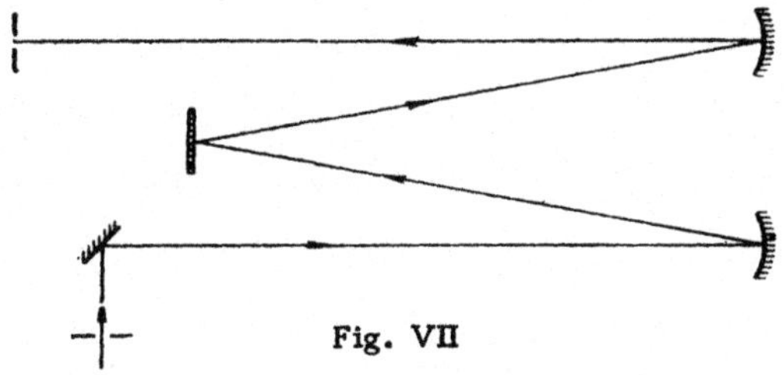

Fig. VII

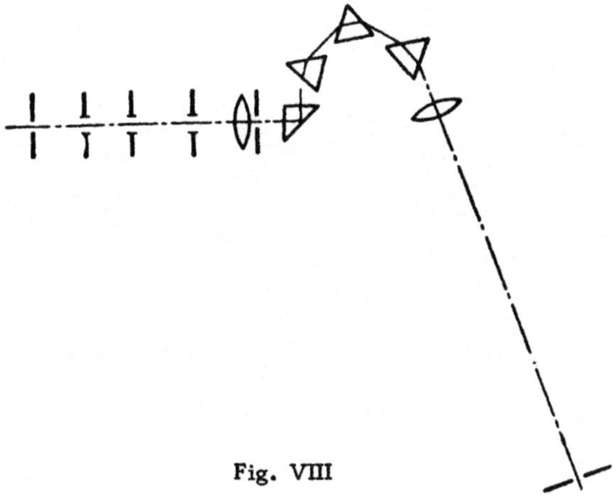

Fig. VIII

APPENDIX IV

Principal types of Soviet photomultipliers used in spectroscopy and their specifications*

Type of PM	Spectral sensitivity region, Å	Maximum sensitivity region, Å	Minimum integral sensitivity, μA/mm	Highest supply voltage, V	Maximum dark current, μA	Photocathode dimensions, mm	Notes
FEU-17A	3200-6000	4000 ± 200	20	900 1400	0.003 0.3	16 x 5	
FEU-18A	2200-6000	4000 ± 200	20	900 1400	0.003 0.3	16 x 5	
FEU-19M	3000-6000	4000 ± 200	10	1100 1400 1900 2600	0.006 0.06 0.6 12	34	
FEU-20	3700-6000	4000 ± 200	20	900	0.008	10 x 5	For portable, pencil-beam equipment
FEU-22	3700-10,000	7500 ± 1000	-	1400	0.2	16 x 5	
FEU-24	3200-8000	4000 ± 200	25	1600 2000	0.3 -	75	
FEU-27	3000-8000	5000 ± 200	30	1100 1500 2000	0.005 - -	25	
FEU-29	3000-6000	4000 ± 200	30	1400 1800 2300	0.03 - -	34	
FEU-31	3000-6000	4000 ± 200	20	850 1300	- 0.5	18	For portable equipment
FEU-33	3000-6000	4000 ± 200	30	2100 2900	1 20	34	For studying processes separated by brief intervals

*For more detailed information on photomultiplier tubes and their characteristics, see [502].

I. Spectrum of Argon. Positive column of high-frequency discharge in argon (a); discharge in a hollow iron cathode. (b). Spectrograph ISP-51 (see Appendix III).

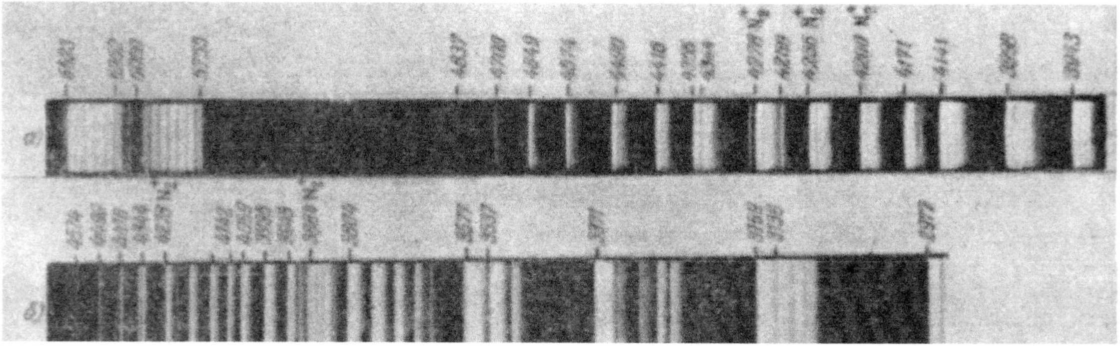

II. Spectrum of Nitrogen. Positive column of high-frequency discharge in air; 1-mm capillary. a) Spectrograph ISP-51; b) Spectrograph ISP-28 (see Appendix III).

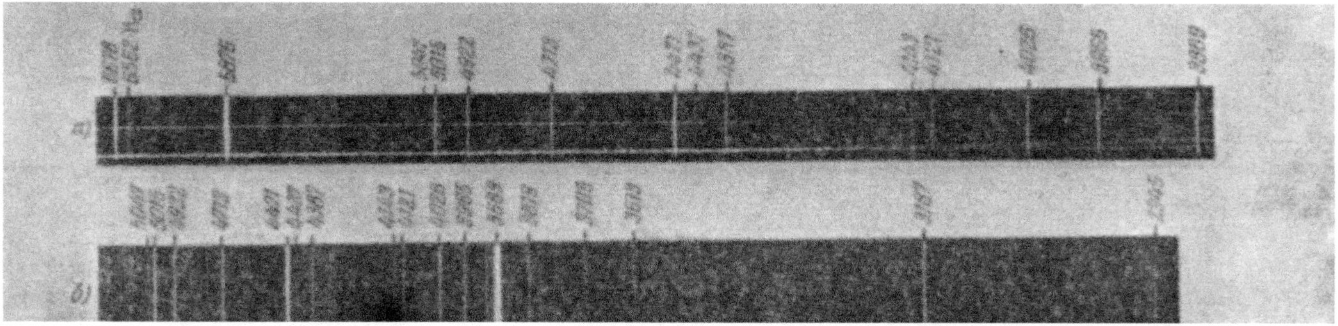

III. Spectrum of Helium. High-frequency discharge in helium. a) Spectrograph ISP-51; b) Spectrograph ISP-28 (see Appendix III).

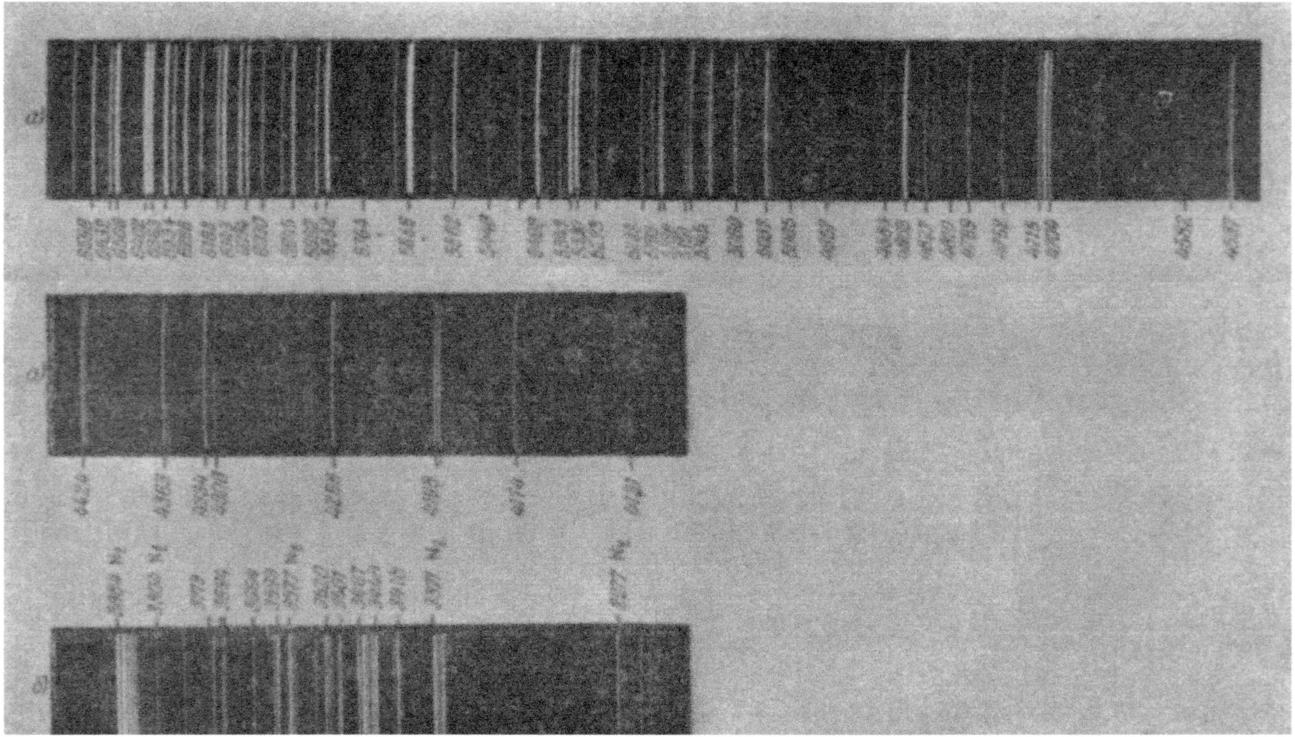

IV. Spectrum of Neon. Positive column of high-frequency discharge in neon; 1-mm capillary, a) Spectrograph ISP-51; b) Spectrograph ISP-28 (see Appendix III). The spectrogram shows traces of nitrogen bands.

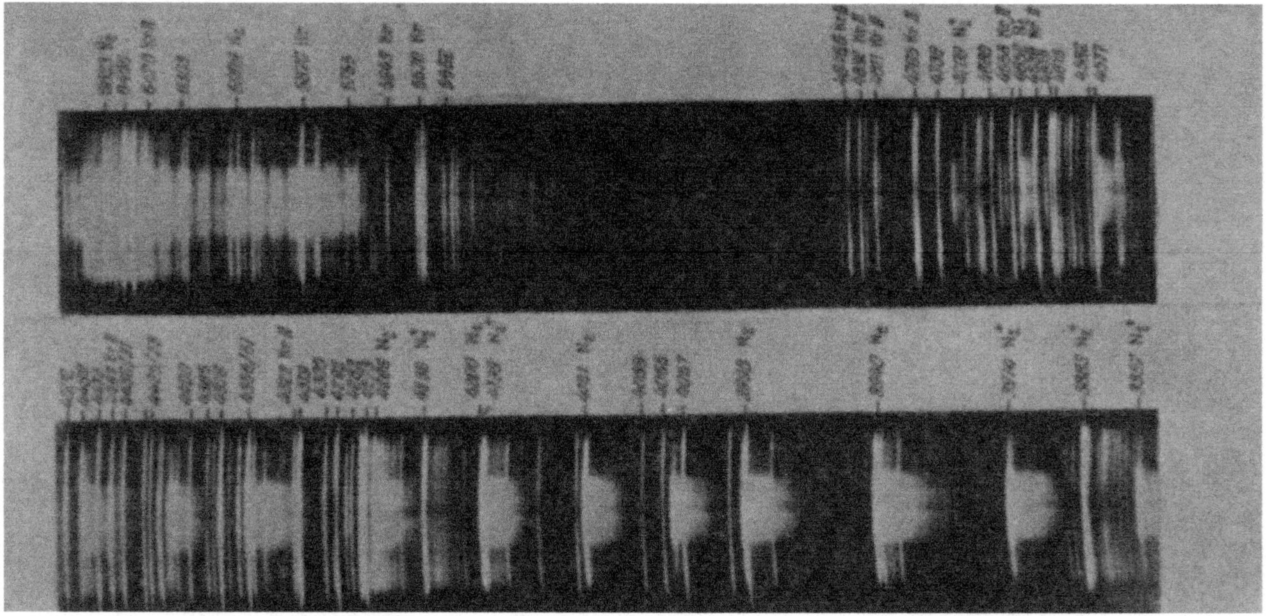

V. Spectrum of Krypton. High-frequency discharge; 1-mm capillary. (The central portion of the plate corresponds to the positive column of the discharge; the edges, to the glow in the vicinity of the external electrodes.) Spectrograph ISP-51 (see Appendix III).

VI. Spectrum of Xenon. Discharge in a hollow iron cathode. Spectrograph ISP-51 (see Appendix III).

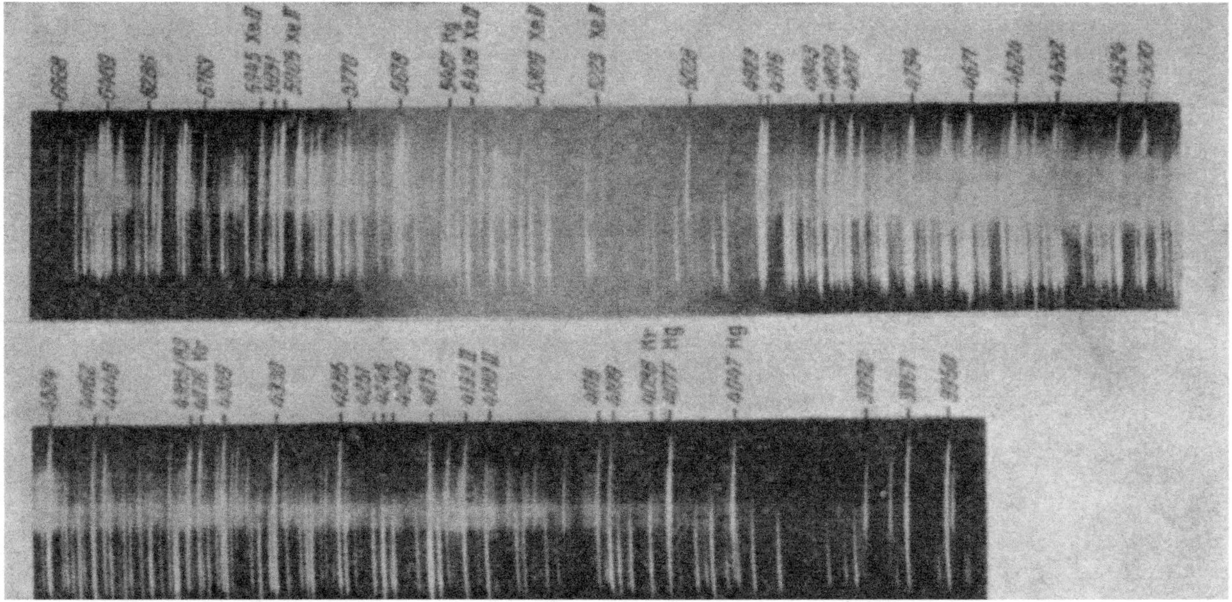

VII. Spectrum of Xenon. High-frequency discharge; 1-mm capillary. (The upper portion of the plate corresponds to excitation in the positive column of the discharge; the lower portion corresponds to excitation in the vicinity of the electrodes.) Spectrograph ISP-51 (see Appendix III).

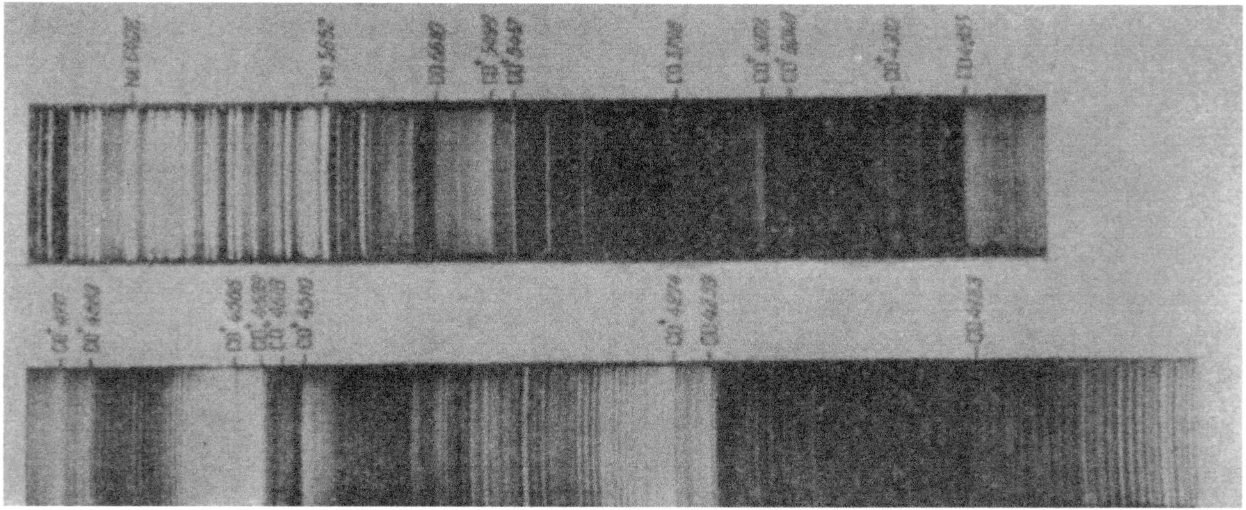

VIII. Spectrum of CO^+. Hollow-cathode discharge in neon containing CO_2. Spectrograph ISP-51 (see Appendix III).

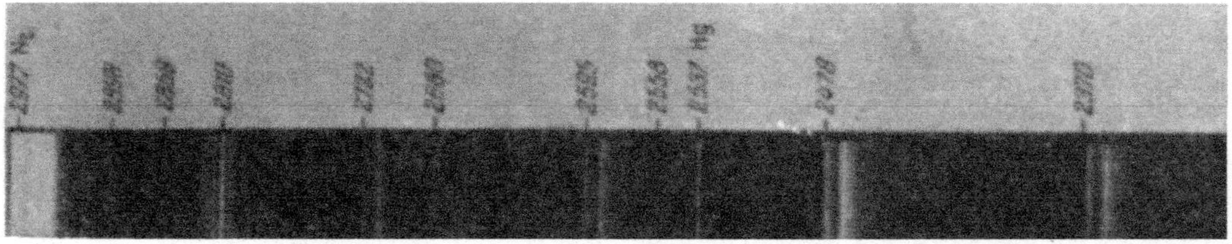

IX. Spectrum of NO. High-frequency discharge in air. Spectrograph ISP-28 (see Appendix III).

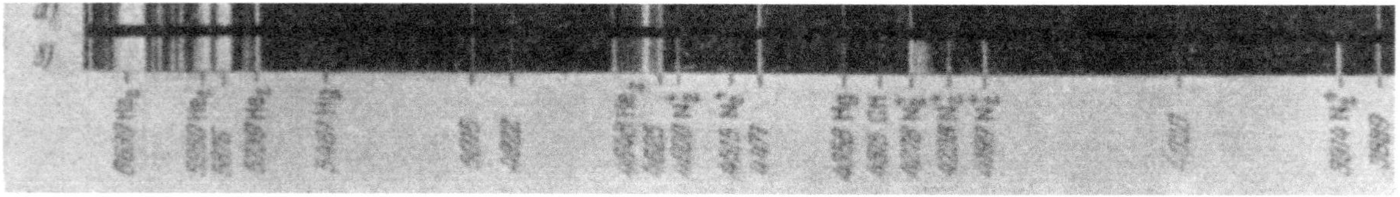

X. Spectra of Helium-Nitrogen mixtures. Positive column of a high-frequency discharge. Pressure about 600 mm, capillary diameter 7 mm. a) N_2 content $6 \times 10^{-4}\%$; b) N_2 content $1 \times 10^{-2}\%$.

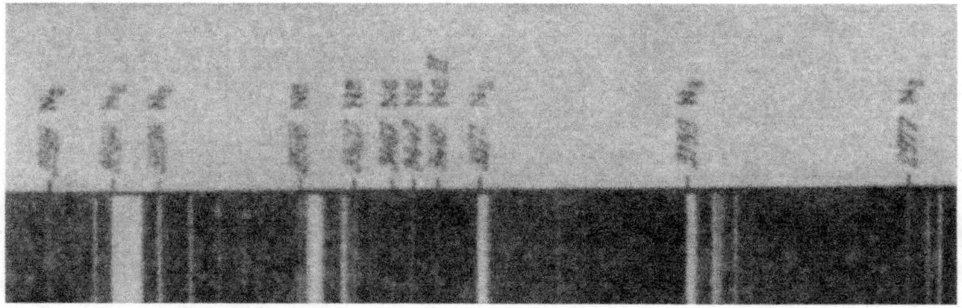

XI. Spectrum of Neon-Nitrogen mixture. Positive column of a high-frequency discharge. Capillary diameter 7 mm. Spectrograph ISP-28 (see Appendix III).

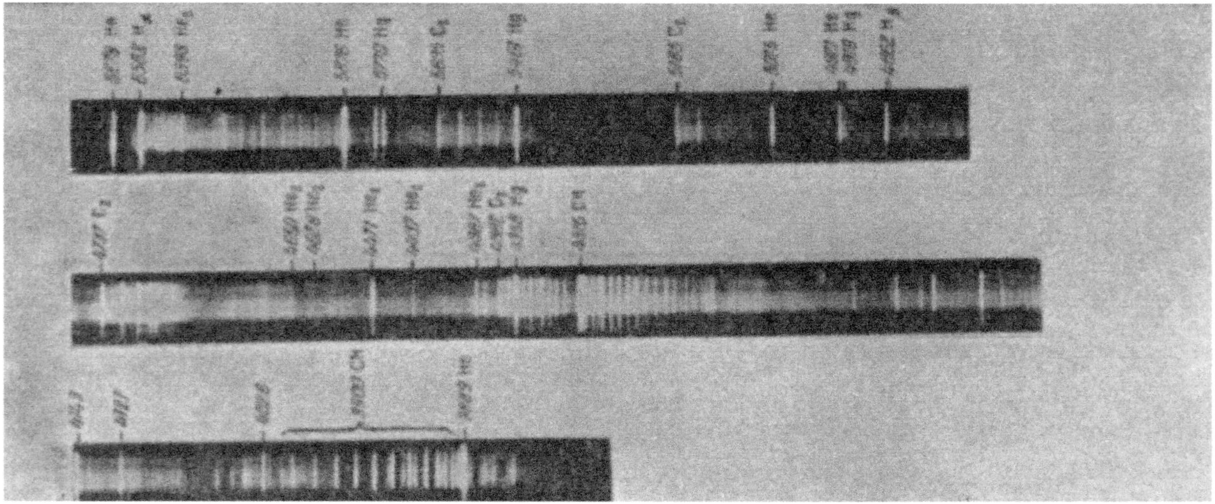

XII. Spectra of Helium-Hydrocarbon mixtures. Positive column of a high-frequency discharge. Capillary diameter 1 mm. Spectrograph ISP-28 (see Appendix III).

REFERENCES

1. Sokolov, V. A. Methods of Gas Analysis. Gostoptekhizdat, 1958.
2. Yeremina, B. G. Gas Analysis. Goskhimizdat, 1955.
3. Guerin, H. Traité de Manipulation et d'Analyse des Gaz, Paris, Masson, 1952.
4. Mullen, P. W. Modern Gas Analysis, New York-London, Interscience Publ., 1955.
5. Zaydel, A. N., V. K. Prokofyev, S. M. Rayskiy, and E. J. Schreider. Tables of Spectral Lines. Fizmatgiz, 1962.
6. Harrison, G. Wavelength Tables of 100,000 Spectrum Lines, New York, 1939.
7. Rusanov, A. K. and N. V. Il'yasova. Atlas of Flame, Arc and Spark Spectra of Elements. Gosgeoltekhizdat, 1958.
8. Kalinin, S. K., A. A. Yavnel', A. I. Alekseyeva, and L. E. Naimark. Atlas of Spectral Lines Obtained with a Quartz Prism Spectrograph. Gostekhizdat, 1952; Gosgeoltekhizdat, 1959.
9. Kalinin, S. K., V. L. Marzuvanov, L. E. Naimark, and K. I. Ismagulov. Atlas of Spectral Lines Obtained with a Glass Prism Spectrograph. Gosgeoltekhizdat, 1956; Publ. Ac. Sci. Kazakh SSR, 1960.
10. Grotrian, W. Graphische Darstellung der Spektren von Atomen und Ionen mit ein, zwei und drei Valenzelektronen, B II, Berlin, J. Springer, 1928.
11. Merril, P. W. Lines of the Chemical Elements in Astronomical Spectra, Washington, 1958, 2 ed.
12. Moore, C. E. Atomic Energy Levels, Vol. I, II, III, Washington, 1949, 1952, 1958.
13. Moore, C. E. A Multiplet Table of Astrophysical Interest, p. 1, 2, Princeton, New Jersey, 1945.
14. Moore, C. E. An Ultraviolet Multiplet Table. Washington, Sect. I, 1950, II, 1952.
15. Frisch, S. E. Optical Spectra of Atoms. Fizmatgiz, 1963.

16. Kondrat'yev, V. N. Atomic and Molecular Structures. Fizmatgiz, 1959.
17. Herzberg, G. Atomic Spectra and Atomic Structure. Russian edition (State Press for Foreign Literature), 1948. English edition, Dover, N. Y., 1944.
18. Shpol'skiy, E. V. Atomic Physics, V. I, II. Gostekhizdat, 1950, 1951.
19. Yel'yashevich, M. A. Atomic and Molecular Spectroscopy. Fizmatgiz, 1962.
20. Pearse, R. W. and A. G. Gaydon. The Identification of Molecular Spectra, London, 1950.
21. Rozhanskiy, D. A. The Physics of Gas Discharge. ONTI, 1937.
22. Kaptsov, N. A. Electric Phenomena in Gases and in Vacuum. Gostekhizdat, 1950.
23. Sena, A. A. Collision of Electrons and Ions with Gas Atoms. Gostekhizdat, 1948.
24. Mitchell, A. C. G. and M. W. Zemansky. Resonance Radiation and Excited Atoms. Russian edition ONTI, 1937.
25. Mandelshtam, S. L. Introduction to Spectral Analysis. Gostekhizdat, 1946.
26. Massey, H. S. W. and E. H. S. Burhop. Electronic and Ionic Impact Phenomena. Russian Edition, 1958. Original: The Clarendon Press, Oxford, 1952.
27. Yakhontova, V. Ye. Vestn. Leningr. Gos. Univ., No. 10, 27 (1959).
28. Hanle, W. Z. Phys. 56, 94 (1929).
29. Lees, J. H. Proc. Roy. Soc., A137, 173 (1932).
30. Steward, D. T. and E. Gabathuler. Proc. Phys. Soc., 74, 473 (1959).
31. St. John, R. M., C. J. Bronco and R. G. Fowler. J. Opt. Soc. Amer., 50, 28 (1960).
32. Schaffernicht, W. Z. Phys., 62, 106 (1930).
33. Frisch, S. E. and I. P. Zapesochnyi. Dokl. Akad. Nauk SSSR, 95, 971 (1954).
34. Herrmann, O. Ann. Phys., 25, 143 (1936).
35. Sin-San Guk and A. M. Devyatov. Vestn. Mosk. Gos. Univ., No. 2, 103 (1957).
36. Smit, J. A. and H. M. Jongerius. Appl. Sci. Res., B5, 59 (1955).
37. Fabrikant, V. A. The Mechanism of Gas Discharge Radiation. In the symposium "Electronic and Ionic Instrumentation," P. V. Timofeyev, ed. Trudy Vsesoyuz. Elektrot. Inst., No. 41, Gosenergoizdat, 1940, p. 236.
38. Fabrikant, V. A. Dokl. Akad. Nauk SSSR, 15, 451 (1937).
39. Frisch, S. E. Vestn. Leningr. Gos. Univ., Mathematical, Physical and Chemical Science Series, No. 8, 129 (1953).

40. Frisch, S. E. Usp. Fiz. Nauk, 43, 512 (1951).
41. Bezold, D. V. Z. angew. Phys., 8, 269 (1956).
42. Smit, Chr. and J. A. Smit. Rev. Universelle Mines, 15, 350 (1959).
43. Tozer, B. A. and J. D. Craggs. J. Electronics and Control, 8, 103 (1960).
44. Ochkur, V. I. Vestn. Leningr. Gos. Univ. Math., Phys. and Chem. Sci. Series, No. 4, 53 (1958).
45. Weinstein, L. A. and G. G. Dolgov. Optika i Spektroskopiya, 7, 3 (1959).
46. Kingston, A. E., B. L. Moiseiwitsch and B. Skinner. Proc. Fourth Intern. Conf. Ion. Phenom. in Gases (Uppsala, 1959), Vol. I, Amsterdam, 1960, p. 236
47. Sluyters, Th. J. M., E. de Haas and J. Kistemaker. Rev. Universelle Mines, 15, 264 (1959).
48. Fedorenko, N. V. Usp. Fiz. Nauk, 68, 481 (1959).
49. Harteck, P., S. Dondes and P. Kentzepis. Z. Phys., 156, 522 (1959).
50. Koch, L. Proc. Fourth Intern. Conf. Ion. Phenom. in Gases (Uppsala, 1959), Vol. 1, Amsterdam, 1960, p. 230.
51. Minnhagen, L. and L. Stigmark. Ark. Fys., 13, 27 (1957); 16, 541 (1960).
52. Kaufman, A. S., T. P. Hughes and R. V. Williams. Proc. Phys. Soc.,76, 17 (1960).
53. Compton, K. T. Phys. Rev., 20, 283 (1922).
54. Compton, K. T. Phil. Mag., 45, 750 (1923).
55. Biberman, L. M. Zh. Exp. Teor. Fiz., 17, 416 (1947).
56. Holstein, T. Phys. Rev., 72, 1212 (1947).
57. Dorgelo, H. B. Physica, 5, 90 (1925).
58. Fabrikant, V. A., F. Butayeva, and I. Tsirg. Zh. Exp. Teor. Fiz., 8, 35 (1938).
59. Barnes, B. T. J. Appl. Phys., 31, 852 (1960).
60. Cayless, M. A. Brit. J. Appl. Phys., 10, 186 (1959).
61. Frisch, S. E. and W. F. Rewald. Optika i Spektroskopiya, 15, (1963).
62. Kagan, Yu. M. and V. M. Zakharova. Zh. Exp. Teor. Fiz., 22, 400 (1952).
63. Frisch, S. E. and Yu. M. Kagan. Ibid., 11, 286 (1941).
64. Meissner, K. W. Ann. Phys., 76, 124 (1925).
65. Nakaya, T. J. Phys. Soc. Japan, 11, 1264 (1956).
66. Meissner, K. W. and W. Graffunder. Ann. Phys., 84, 1009 (1927).
67. Kenty, C. Phys. Rev., 80, 95 (1950).
68. Penning, F. M. Phil. Mag., 11, 961 (1931).
69. Kolesnikov, V. N. and N. N. Sobolev. Radiotekhnika i Elektronika, 4, 1286 (1959).

70. Sobolev, N. N., V. N. Kolesnikov and V. F. Kitaeva. Proc. IV Intern. Conf. Ion. Phenom. in Gases (Uppsala, 1959), Vol. I, Amsterdam, 1960, p. 370.
71. Langmuir, J. and H. Mott-Smith. General Electric Rev., 27, 449, 538, 616, 762, 810 (1924).
72. Killian, T. J. Phys. Rev., 35, 1238 (1930).
73. Kagan, Yu. M. and R. I. Ligushchenko. Zh. Tekh. Fiz., 31, 445 (1961).
74. Boyd, R. L. F. and N. D. Twiddy. Proc. Roy Soc., A250, 53 (1959).
74a. Vorob'yeva, N. A., Yu. M. Kagan and V. M. Milenin. Zh. Tekh. Fiz., 33 (1963).
75. Klyarfel'd, B. N. The Positive Column of Gas Discharge as a Source of Light. In the symposium "Electronic and Ionic Instrumentation," P. V. Timofeyev, ed. Trudy Vsesoyuz. Elektrot. Inst., No. 41, Gosenergoizdat, 1940, p. 166.
76. Johnson, E. O. and L. Malter. Phys. Rev., 80, 58 (1950).
77. Biberman, L. and B. Panin. Zh. Tekh. Fiz., 21, 12 (1951).
78. Tverdokhlebov, V. I. Ibid., 27, 753 (1957).
79. Kagan, Yu. M., V. I. Perel' and P. O. Ripatti. Vestn. Leningr. Gos. Univ., Phys., Chem. and Math. Sci. Series, No. 8, 129 (1955).
80. Malyshev, G. M. and V. L. Fedorov. Dokl. Akad. Nauk SSSR, 92, 269 (1953).
81. Richelmann, B. Rev. Sci. Instrum., 30, 593 (1959).
82. Waymouth, J. F. J. Appl. Phys., 30, 1404 (1959).
83. Levitskiy, S. M. and I. P. Shashurin. Zh. Tekh. Fiz., 31, 436 (1961).
84. Fetz, H. and H. Oechsner. Z. angew. Phys., 12, 250 (1960).
85. Dzerpetov, Kh. A. and G. M. Pateyuk. Zh. Eksp. Teor. Fiz., 28, 343 (1955).
86. Levitskiy, S. M. and I. P. Shashurin. Radiotekhnika i Elektronika, 4, 1238 (1959).
87. Avramov, V. V. and Kh. A. Dzerpetov. Vestn. Mosk. Gos. Univ., No. 3, 55 (1960).
88. Shukhtin, A. M. Optika i Spektroskopiya, 7, 838 (1959).
89. Alpher, R. A. and D. R. White. Phys. of Fluids, 1, 452 (1958).
90. Seeliger, R. and R. Hirchert. Ann. Phys., 11, 817 (1931).
91. Zaytsev, A. A. and Ye. I. Yankovskaya. Dokl. Akad. Nauk SSSR, 29, 562 (1940).
92. Vilenskaya, T. V. and A. S. Makarova. Izv. Vyssh. Ucheb. Zav., Fizika, No. 6, 102 (1959).
93. Kasabov, G. and V. A. Fabrikant. Proc. of the X All-Union Conference on Spectroscopy, Lvov, Vol. II, 51 (1958).
94. Zakharova, V. M. and Yu. M. Kagan. Optika i Spektroskopiya, 1, 627 (1956).

95. Mohler, F. L. J. Res. Nat. Bur. Stand., 21, 696 (1938).
96. Klyarfel'd, B. N. Zh. Tekh. Fiz., 8, 410 (1938).
97. Karelina, N. A. Collected Papers on Technical Physics, 1948, p. 167.
98. Engelke, B. A. Z. Phys., 158, 422 (1960).
99. Bochkova, O. P., L. P. Razumovskaya and S. E. Frisch. Optika i Spektroskopiya, 11, 697 (1961).
100. Zaytsev, A. A. Vestn. Mosk. Gos. Univ., No. 10, 69 (1947).
101. Zaytsev, A. A. Zh. Exp. Teor. Fiz., 8, 569 (1938).
102. Dorgelo, H. B., H. Alting and C. J. Boers. Physica, 2, 959 (1935).
103. Malyshev, G. M. and B. V. Pavlov. Vestn. Leningr. Gos. Univ., No. 16, 32 (1957).
104. Morgenroth, H. Ann. Phys., 3, 373 (1959).
105. Engel, A. and M. Steenbeck. The Physical Phenomena and Experimental Techniques Relating to Electrical Discharge in Gases. Russian Edition ONTI, v. I, (1935); v. II, (1936).
106. Penkin, N. P. Optika i Spektroskopiya, 2, 545 (1957).
107. Penning, F. M. Electrical Discharges in Gases. Russian Edition, 1960. Translated by A. F. Montgomery, Macmillan, N. Y., 1957.
108. Pahl, M. and U. Weimer. Z. Naturforsch., 12a, 926 (1957).
109. Rozhdestvenskiy, D. S. Zh. Russ. Fiz. Khim. Obshch., 44, 395 (1912).
110. Rozhdestvenskiy, D. S. Studies in Anomalous Dispersion of Metal Vapors. S. E. Frisch, editor. 1951.
111. Ladenburg, R. Usp. Fiz. Nauk, 14, 721 (1934).
112. Penkin, N. P. Izv. Akad. Nauk SSSR, Physics Series, 11, 217 (1947).
113. Kagan, Yu. M. and N. P. Penkin. Ibid., 14, 721 (1950).
114. Penkin, N. P. and M. N. Palladin. Ibid., 19, 16 (1955).
115. Shukhtin, A. M. and V. S. Yegorov. Vestn. Leningr. Gos. Univ., No. 16, 61 (1959).
116. Ladenburg, R. and F. Reiche. Ann. Phys., 42, 181 (1913).
117. Ladenburg, R. and S. Levy. Z. Phys., 65, 189 (1930).
118. Fabrikant, V. A. and V. Pul'ver. Dokl. Akad. Nauk SSSR, 3, 435 (1934).
119. Fabrikant, V. A., F. A. Butayeva and I. Tsirg. Zh. Exp. Teor. Fiz., 7, 733 (1937).
120. Ladenburg, R. and S. Levy. Z. Phys., 88, 461 (1934).
121. Panevkin, K. I. Zh. Exp. Teor. Fiz., 9, 1007 (1939).
122. Frisch, S. E. Izv. Akad. Nauk SSSR, Physics Series, 14, 711 (1950).
123. Frisch, S. E. and I. P. Bogdanova. Symposium in memory of S. I. Vavilov. Published by Acad. Sci. USSR, 1952, p. 220.
124. Frisch, S. E. and O. P. Bochkova. The 13th Conference on

Spectroscopy, 1960; Physical Problems in Spectroscopy, I, 142 (1962).
125. Frisch, S. E. and O. P. Bochkova. Vestn. Leningr. Gos. Univ., Chem. Phys. Series, No. 16, 40 (1961).
126. Harrison, J. A. Proc. Phys. Soc., 73, 841 (1959).
127. Biberman, L. M. Optika i Spektroskopiya, 3, 397 (1957).
128. Filippov, A. N. Trudy Gos. Opt. Inst., 8, No. 85 (1932).
129. Kohn, H. Phys. Z. S., 29, 49 (1928); 33, 957 (1932).
130. Mohler, F. L. J. Res. Nat. Bur. Stand., 16, 227 (1936).
131. Penkin, N. P. and A. M. Shukhtin. Izv. Akad. Nauk SSSR, Phys. Series, 12, 376 (1948).
132. Penkin, N. P. and A. M. Shukhtin. Uch. Zap. Leningr. Gos. Univ., No. 7, 28 (1949).
133. Kopfermann, H. and R. Ladenburg. Z. Phys., 65, 167 (1930).
134. Levy, S. Z. Phys., 72, 578 (1931).
135. Schön, K. Ann. Phys., 28, 649 (1937).
136. Kruse, M. Z. Phys., 109, 312 (1938).
137. Krebs, K. Z. Phys., 101, 604 (1936).
138. Bogdanova, I. P. Vestn. Leningr. Gos. Univ., Phys. and Chem. Series, No. 22, 41 (1956).
139. Herzberg, G. Z. Phys., 146, 269 (1956).
140. Turkin, Yu. I. Optika i Spektroskopiya, 2, 290 (1957).
141. Turkin, Yu. I. Ibid., 7, 10 (1959).
142. Frisch, S. E. Usp. Fiz. Nauk, 61, 461 (1957).
143. Zakharova, V. M. and Yu. M. Kagan, Vestn. Leningr. Gos. Univ., No. 16, 44 (1959).
144. Kagan, Yu. M. and V. I. Perel'. Optika i Spektroskopiya, 2, 298 (1957).
145. Condell, W. J. J. Opt. Soc. Amer., 50, 610 (1960).
146. Bogdanova, I. P. and Ch'eng-Gi-Thek. Optika i Spektroskopiya, 2, 681 (1957).
147. Bogdanova, I. P. and S. E. Frisch. Studies in Experimental and Theoretical Physics. In symposium dedicated to the memory of G. S. Landsberg, 1959, p. 275.
148. Schneider, F. Ann. Phys., 18, 297 (1956).
149. Tsui-Fang and J. D. Craggs. J. Electronics Control, 4, 385 (1958).
150. Harrison, J. A and J. D. Craggs. Ibid., 4, 289 (1958).
151. Dutta, A. K and U. Ganntayet. J. Opt. Soc. Amer., 46, 1046 (1956).
152. Yakhontova, V. Ye. and M. I. Kliot-Dashinskiy. Optika i Spektroskopiya, 7, 446 (1959).
153. Little, P. F. and A. von Engel. Proc. Roy. Soc., A224, 209 (1954).
154. Phelps, A. V. Phys. Rev., 110, 1362 (1958).
155. Gromov, V. A. Optika i Spektroskopiya, 1, 334 (1956).

156. Kunisz, M. D. and J. Séguier. J. Phys. Rad., 21, 527 (1960).
157. Baker, M. R., S. J. Adelstein and B. L. Vallee. J. Opt. Soc. Amer., 46, 138 (1956).
158. Lomonosova, L. S., A. V. Nedospasov and A. Ye. Novik. Proc. of the 10th Conference on Spectroscopy, Lvov, vol. II, 1958, p. 201.
159. Frisch, S. E. and A. A. Ferchmin. Phys. Zeitschr. Sow. Union, 9, 466 (1936).
160. Frisch, S. E., and E. K. Kraulinya. Dokl. Akad. Nauk SSSR, 101, 837 (1955).
161. Frisch, S. E., E. K. Kraulinya and I. P. Bogdanova. Proc. of the 10th Conference on Spectroscopy, Lvov, vol. II, 1958, p. 54.
162. Sychev, V. P. Optika i Spektroskopiya, 6, 419 (1959).
163. Schmidt, K. Z. Naturforsch., 11a, 1023 (1956).
164. Hanle, W. and K. Schmidt. Ann. Phys., 20, 173 (1957).
165. Hagen, S. and R. Ritschl. Spectr. Acta. 14, 477 (1959).
166. Friedl, W. and S. Hagen. Optik, 15, 686 (1958).
167. Friedrich, H. and R. Seiwert. Ann. Phys., 20, 215 (1957).
168. Friedl, W. Z. Naturforsch., 15a, 398 (1960).
169. Javan, A., W. R. Bennet and D. R. Herriott. Phys. Rev. Lett., 6, 106 (1961).
170. Addink, N. W. H. Spectr. Acta., 14, 349 (1959).
171. Leb, L. Basic Processes Involving Electric Discharges in Gases. Gostekhizdat, 1950.
172. Radiophysical Electronics. N. A. Kaptsov, ed. Moscow Univ. Press, 1960.
173. Engel, A. Ionized Gases. Fizmatgiz, 1959.
174. Meek, J. D. and J. D. Craggs. Electrical Breakdown in Gases. The Clarendon Press, Oxford, 1953. Russian edition 1960.
175. Jenkins, J. and T. B. Jones. J. Appl. Phys., 28, 663 (1957).
176. Boyle, W. S. and F. E. Haworth. Phys. Rev., 101, 935 (1956).
177. Schlumbohm, H. Z. Phys., 159, 212 (1960).
178. Neu, H. Z. Phys., 154, 423 (1959); 155, 77 (1959).
179. Blodget, K. B. and T. A. Vanderslice. J. Appl. Phys., 31, 1017 (1960).
180. Bartholomew, C. Y. and A. R. La Padula. J. Appl. Phys., 31, 445 (1960).
181. Hall, L. D. Rev. Sci. Instrum., 29, 367 (1958).
182. Holland, L. D. Ibid., 36, 105 (1959).
183. Strotzer, G. Z. angew. Phys., 10, 207 (1958); 11, 223 (1959).
184. Carter, G. Vacuum, 9, 190 (1959).
185. Kagan, Yu. M. and V. I. Perel'. Optika i Spektroskopiya, 4, 285 (1958).
186. Rokhlin, G. N. Zh. Exp. Teor. Fiz., 9, 804 (1939).
187. Fataliyev, Kh. M. Ibid., 9, 813 (1939).
188. Slutskin, A. A. Zh. Tekh. Fiz., 7, 1862 (1937).

189. Davies, L. W. Proc. Phys. Soc., B66, 898 (1953).
190. Tardy, O. and R. Lennuier. J. Phys. Rad., 19, 75A (1958).
191. Klyarfel'd, B. N. Zh. Exp. Teor. Fiz., 22, 66 (1952).
192. Wojaczek, K. Ann. Phys., 3, 37 (1959).
193. Nedospasov, A. V. Zh. Tekh. Fiz., 28, 173 (1958).
194. Ruttenauer, A. Z. Phys., 10, 269 (1922).
195. Skaupy, F. Z. Phys., 2, 213, 1920; Verh. d. Deut. Phys. Ges., 18, 230 (1916); Skaupy, F. and F. Bobek. Z. f. techn. Phys., 6, 284 (1925).
196. Vygodskiy, Ya. S. and B. N. Klyarfel'd. Zh. Tekh. Fiz., 3, 610 (1933).
197. Druyvesteyn, M. J., Physica, 2, 255 (1935).
198. Riesz, R. and G. H. Dieke. J. Appl. Phys., 25, 196 (1954).
199. Krysmanski, K. H. Ann. Phys., 2, 263 (1958).
200. Groth, W. and P. Harteck. Naturwissenschaften, 27, 390 (1939).
201. Beckey, H. D., W. E. Groth and K. H. Welge. Z. Naturforsch., 8a, 556 (1953).
202. Matveyeva, N. A. Zav. Lab., 24, 746 (1958).
203. Matveyeva, N. A. Vestn. Leningr. Gos. Univ., No. 4, 11 (1959).
204. Frisch, S. E. and N. A. Matveyeva. Dokl. Akad. Nauk SSSR, 122, 375 (1958).
205. Loeb, L. B. J. Appl. Phys., 29, 1369 (1958).
206. Kagan, Yu. M. and G. M. Malyshev. Zh. Tekh. Fiz., 23, 950 (1953).
207. Paschen, F. Ann. Phys., 50, 901 (1916).
208. Tolansky, S. High Resolution Spectroscopy. Russian Edition, 1955.
209. Zaydel', A. N., N. I. Kaliteyevskiy, L. V. Lipis, and M. P. Chayka. Emission Analysis of Atomic Materials. Fizmatgiz, 1960.
210. Jones, W. G. and A. Walsh. Spectr. Acta., 16, 249 (1960).
211. Veith, W. Naturwissenschaften, 42, 40 (1955).
212. Lompe, A., R. Seeliger and E. Wolter. Ann. Phys., 36, 9 (1939).
213. Badareu, E., I. Popescu and I. Iova. Ann. Phys., 5, 308 (1960); Rev. de Physique, 5, 287 (1960).
214. Holstein, T. Phys. Rev., 70, 367 (1946).
215. Margenau, H. Phys. Rev., 73, 297, 309, 326 (1948); Phys. Rev., 74, 706 (1948).
216. Hartman, L. M. Phys. Rev., 73, 316 (1948).
217. Hiedemann, E. Ann. Phys., 85, 649 (1928).
218. Lodge, J. I. and R. W. Stewart. Canad. J. Res., A26, 205 (1948).
219. Levitskiy, S. M. Zh. Tekh. Fiz., 27, 1001 (1957).
220. Banerji, D. and R. Ganguli. Phil. Mag., 15, 676 (1933); Banerji, D. and D. Bhattacharya. Phil. Mag., 17, 313 (1934).
221. Bay, Z. and W. Steiner. Z. phys. Chem., B3, 149 (1929).

222. Pateyuk, G. M. Zh. Exp. Teor. Fiz., 30, 12 (1956).
223. Levitskiy, S. M. Zh. Tekh. Fiz., 27, 970 (1957).
224. Salmon, J. Ann. Physique, 2, 827 (1957).
225. Gill, E. W. and A. von Engel. Proc. Roy. Soc., A192, 446 (1948).
226. Beck, H. Z. Phys., 97, 335 (1935).
227. Dzerpetov, Kh. A. and G. M. Pateyuk. Zh. Exp. Teor. Fiz., 28, 343 (1955).
228. Kojima, S. K. Takayma and A. Shimauchi. J. Phys. Soc. Japan, 8, 55 (1953).
229. Bhatt, N. B. Proc. Ind. Acad. Sci., 1, 891 (1935).
230. Nöller, W. Z. Phys., 110, 320 (1938).
231. Fox, G. W. and C. H. Bachman. Phys. Rev., 50, 939 (1936).
232. Robertson, J. K. and R. H. Hay. Canad. J. Res., 14, 201 (1936).
233. Hayes, K. I., J. G. Winans and C. Warren. Phys. Rev., 68, 98(A) (1945).
234. Tawde, N. R. and G. K. Mehta. Proc. Ind. Acad. Sci., 23, 67 (1945); Phys. Rev., 69, 245 (1946).
235. Vagner, S. D., Yu. M. Kagan and Ye. V. Romanova. Vestn. Leningr. Gos. Univ., No. 10, 15 (1958).
236. Vavilin, Ye. I., S. D. Vagner and A. M. Drukman. Zh. Tekh. Fiz., 29, 1263 (1959).
237. Razumovskaya, L. P. and O. P. Bochkova. Optika i Spektroskopiya, 9, 271 (1960).
238. Golant, V. Ye. Usp. Fiz. Nauk, 65, 39 (1958).
239. Bakakina, V. V. Optika i Spektroskopiya, 9, 784 (1960).
240. Meggers, W. F. and F. O. Westfall. J. Res. Nat. Bur. Stand., 44, 447 (1950).
241. Vaschman, A. A., L. V. Lipis and N. A. Teterina. Optika i Spektroskopiya, 6, 260 (1959).
242. Mitsuk, V. Ye., M. D. Koz'minykh and I. V. Talalayeva. Izv. Akad. Nauk SSSR, Phys. Ser., 23, 1031 (1959).
243. Jacobsen, E. and G. R. Harrison. J. Opt. Soc. Amer., 39, 1054 (1949).
244. Forrester, A. T., R. A. Gundmundsen and P. O. Johnson. J. Opt. Soc. Amer., 46, 339 (1956).
245. Meggers, W. F. J. Opt. Soc. Amer., 38, 7 (1948).
246. Cabannes, F. Ann. Physique, 10, 1026 (1955).
247. Ham, N. S. and A. Walsh. Spectrochim. Acta, 12, 88 (1958).
248. Thomson, J. Phil. Mag., 4, 1128 (1927).
249. Mierdel, G. Ann. Phys., 85, 612 (1928).
250. Wachsmuth, R. Ann. Phys., 39, 611 (1912); Wachsmuth, R. and B. Winaver. Ann. Phys., 42, 585 (1913).
251. Strauss, H. J. Ann. Phys., 1, 281 (1958).
252. Davydov, B. I. Symposium on Plasma Physics and Controlled Thermonuclear Reactions. Acad. Sci. USSR, 1, 89 (1958).

253. Birkkhoff, G. Z. angew. Phys., 10, 204 (1958).
254. Esclangon, F. Ann. Physique, 1, 267, (1934).
255. Bloch, L. and E. Bloch. J. de Phys., 4, 333 (1923).
256. Dejardin, G. Ann. Physique, 2, 241 (1924).
257. Bloch, L., E. Bloch and G. Dejardin. J. de Phys., 7, 129 (1926).
258. Bloch, L. and E. Bloch. Ann. Physique, 13, 233 (1930).
259. Goudet, G., P. Herreng and G. Niff. Compt. Rendus Acad. Sci., 214, 62 (1942).
260. Roig, J. and A. Buret. Ibid., 230, 1394 (1950).
261. Ewest, H. Z. techn. Phys., 20, 294 (1939).
262. Drewell, P. Elektrotech. Zeitsch., 60, 1335 (1939).
263. Bogdanov, S. Ya. and K. S. Wulfson. Dokl. Akad. Nauk SSSR, 30, 309 (1941).
264. Wulfson, K. S. Izv. Akad. Nauk SSSR, Phys. Ser., 9, 239 (1945); Electrichestvo, No. 11, 16 (1946).
265. Shukhtin, A. M. and V. S. Yegorov. Optika i Spektroskopiya, 2, 543 (1957); Shukhtin, A. M., V. S. Yegorov and G. K. Tumakayev. Ibid., 8, 423 (1960).
266. L. Ch. Usp. Fiz. Nauk, 42, 500 (1950).
267. Karyakina, A. V. Ibid., 53, 413 (1954).
268. Murphy, P. M. and E. H. Edgerton. J. Appl. Phys., 12, 848 (1941).
269. Hoyt, G. D. and W. W. McCormick. J. Opt. Soc. Amer., 40, 658 (1950).
270. Sittner, W. R. and E. R. Peck. J. Opt. Soc. Amer., 39, 474 (1949).
271. Bochkova, O. P. and E. J. Schreider. Izv. Akad. Nauk SSSR, Phys. Ser., 19, 75 (1955).
272. Laporte, M. Les Lampes à Eclaire Lumiére Blanche et Leurs Applications, Paris, 1949.
273. Bardocz, A. Appl. Spectr., 11, 167 (1957).
274. McNobbs, J. Spectr. Acta, 14, 939 (1959).
275. Spä̈th, H. and H. Krempl. Z. angew. Phys., 12, 8 (1960).
276. Dushman, S. Scientific Foundations of Vacuum Engineering. John Wiley & Sons, N. Y., 1962. Russian Edition 1950.
277. Chmutov, K. V. Physico-chemical Research Techniques. Goskhimizdat, 1954.
278. Angerer, E. Laboratory Techniques. Russian Edition. Fizmatgiz, 1962.
279. Strong, D. Procedures in Experimental Physics. Prentice Hall, N. J., 1938, Russian Edition, 1948.
280. Korolev, B. I. Foundations of Vacuum Engineering. Gosenergoizdat, 1958.
281. Groszkowski, J. High Vacuum Technology. Russian edition, 1957.

282. Espe, W. and M. Knoll. Technology of Electrovacuum Materials. Russian Edition, 1939.
283. Yarwood, J. High Vacuum Engineering Techniques, John Wiley & Sons, N. Y., 1955. Russian Edition, Gosenergoizdat, 1960.
284. Jansen, C. G. J. and A. Venema. Vacuum, 9, 219 (1959).
285. Shishakov, N. A. Zh. Prikl. Fiz., 3, 279 (1926).
286. Terrien, J. Rev. Optique, 38, 29 (1959).
287. Veis, S. Vacuum, 9, 186 (1959).
288. Drawin, H. W. Vacuum Technik, 7, 177 (1958).
289. Mirtov, B. A. and A. A. Boykov. Zav. Lab. 21, 871 (1955).
290. Zaydel', A. N., A. A. Petrov and G. V. Veynberg. Hydrogen Determination in Metals by Mass Spectroscopy. Leningr. Univ. Press, 1957.
291. Paty, L. Vacuum, 7, 8, 80 (1957, 1958).
292. Astrov, D. N. and A. V. Voronel'. Fiz. i Tekh. Eksperim., No. 3, 149 (1960).
293. Vidro, G. I., B. D. Luft and Yu. V. Matorin. Proc. of the 10th All-Union Conference on Spectroscopy, Lvov, Vol. II, 1958, p. 212.
294. Kotlik, L. L. and L. S. Lomonosova. Zav. Lab., 25, 1482 (1959).
295. Hansen, N. R. J. Sci. Instrum., 32, 75 (1955).
296. Roberts, R. M. and J. J. Madison. Anal. Chem., 29, 1555 (1957).
297. McKelvey, J. M. and H. E. Hoelscher. Anal. Chem., 29, 123 (1957).
298. Hersch, P. and J. E. Whittle. J. Sci. Instrum., 35, 32 (1958).
299. Ageykin, D. I., E. L. Itskovitch and S. P. Mikhaylov. Zav. Lab., 23, 492 (1957).
300. Brunauer, S. The Adsorption of Gases and Vapors. Princeton University Press, N. Y., 1943. Russian Edition, 1948.
301. Manley, J. J. Proc. Phys. Soc., 36, 288 (1924).
302. Curry, J. R. J. Phys. Chem., 35, 859 (1931).
303. Frank, H. S. J. Phys. Chem., 33, 970 (1929).
303a. Morgulis, N. D., Yu. G. Ptushinskiy and B. A. Chuykov. Dokl. Akad. Nauk SSSR, 128, 930 (1959).
304. Hickmott, T. W. J. Appl. Phys., 31, 128 (1960).
305. Blodgett, K. B. J. Chem. Phys., 29, 39 (1958).
306. Whetten, N. R. and J. R. Young. Rev. Sci. Instrum., 30, 472 (1959).
307. Bennett, W. R. Rev. Sci. Instrum., 28, 1092 (1957).
308. Fastovskiy, V. G. and A. Ye. Rovinskiy. Zh. Prikl. Khimii, 33, 1641 (1960).
309. Krummer, J. T. Appl. Spectroscopy, 12, 155 (1958).
310. Ustinov, V. B., O. P. Bochkova and L. P. Razumovskaya. Zav. Lab., 26, 621 (1960).

311. Zaydel', A. N., A. A. Petrov and V. B. Ustinov. Ibid., 27, 904 (1961).
312. Vernotte, M. Groupement pour l'Avancement des Méthodes d'Analyse Spectrographique des Produits Metallurgiques, XIX Congres, Paris, 1956, p. 357.
313. Frisch, S. E. Spectroscopic Techniques. Leningr. Univ. Press, 1936.
314. Lomonosova, L. S. and O. B. Fal'kova. Spectral Analysis. Gostekhizdat, 1958.
315. Prokof'yev, V. K. Photographic Methods in Quantitative Spectral Analysis of Metals and Alloys, Parts I, II, Gostekhizdat, 1951.
316. Sawyer, R. A. Experimental Spectroscopy. Prentice Hall, N. Y., 1951, Russian Edition, 1953.
317. Harrison, George R., Richard C. Lord and John R. Loofbourow. Practical Spectroscopy. Prentice Hall, N. Y., 1948, Russian Edition, 1950.
318. Jacquinot, P. Usp. Fiz. Nauk, 72, 799 (1960).
319. Vinogradova, A. K. and L. M. Ivantsov. Optika i Spektroskopiya, 6, 829 (1959).
320. Boillet, P. Rev. Opt., 39, 47, 147 (1960).
321. The Soviet Tinted Glass Catalog. Oborongiz, 1960.
322. Christiansen, C. Ann. Physik und Chemie, 23, 298 (1884); 24, 439 (1885).
323. Korolev, F. A. and A. Yu. Klement'yeva. Dokl. Akad. Nauk SSSR, 94, 1025 (1954).
324. Lyot, B. Compt. Rendus Acad. Sci., 197, 1593 (1933).
325. Baranov, S. and N. Melankholin. Zh. Tekh. Fiz., 15, 89 (1945).
326. Geffcken, W. Angew. Chem., 60, 1 (1948).
327. Korolev, F. A. Izv. Akad. Nauk SSSR, Phys. Ser., 11, 448 (1947).
328. Krylova, T. N. Optika i Spektroskopiya, 6, 784 (1959).
329. Weigert, F. and H. Staude, Z. Phys. Chem., A130, 607 (1927).
330. Kohn, H. and K. Fragstein. Phys. Zeitschr., 33, 929 (1932).
331. Barnes, R. B. and L. G. Bonner. Phys. Rev., 49, 732 (1936).
332. Sethi, N. K. Proc. Ind. Assoc. Cult. Sci., 6, 121 (1921).
333. Sogani, C. M. Phil. Mag., 1, 321 (1926).
334. Raman, C. V. Proc. Ind. Acad. Sci., A29, 381 (1949).
335. Hillvarg, L. V., G. I. Distler and Ye. A. Makarova. Dokl. Akad. Nauk SSSR, 100, 1067 (1955).
336. Korolev, F. A. Vestn. Mosk. Gosud. Univ., No. 3, 97 (1958).
337. Schröder, H. Z. angew. Phys., 3, 53 (1951).
338. Dufour, Ch. J. Phys. Radium, 11, 327 (1950); Rev. Opt., 31, 1 (1952).

339. Polster, H. D. J. Opt. Soc. Amer., 42, 21 (1952).
340. Korolev, F. A. and A. Yu. Klement'yeva. Vestn. Mosk. Gosud. Univ., No. 3, 65 (1957).
341. Krylova, T. N. Optika i Spektroskopiya, 4, 217 (1958).
342. Krylova, T. N. Ibid., 6, 784 (1959).
343. Sokolova, R. S. and T. N. Krylova. Ibid., 6, 788 (1959).
344. Kliment'yeva, L. Yu. Vestn. Mosk. Gosud. Univ., No. 1, 23 (1960).
345. Heavlens, O. S., J. Ring and S. D. Smith. Spectrochim. Acta, 10, 179 (1957).
346. Sinel'nikov, K. D. and N. A. Vlasenko. Zh. Tekh. Fiz., 26, 96 (1956).
347. Speranskaya, N. I. and G. N. Rautiyan. Ibid., 22, 620 (1952).
348. Chechik, N. O., S. M. Feinstein and T. M. Lifshits. Electronic Multipliers. Gostekhizdat, Moscow, 1957, 2nd ed.
349. Rodionov, S. F., A. L. Osherovich, Ye. N. Palova, and L. M. Fishkova. Zh. Tekh. Fiz., 19, 184 (1949).
350. Description of the PS-381 Instrument. State All-Union Plant, 1956.
351. Description of the PS-382 Instrument. Ibid., 1956.
352. Description of the FEP-1 Instrument. Ibid., 1959.
353. Abramson, I. S. Zav. Lab., 20, 168 (1954).
354. Saunderson, J. L., V. J. Caldecourt and E. W. Peterson. J. Opt. Soc. Amer., 35, 681 (1945).
355. Carpenter, R. O. B., E. DuBois, and Sterner. J. Opt. Soc. Amer., 37, 707 (1947).
356. Oldenberg, O. and H. P. Broida. J. Opt. Soc. Amer., 40, 381 (1950).
357. Krumbein, A. D., F. A. Grant and A. L. Ward. J. Opt. Soc. Amer., 42, 277 (L) (1952).
358. Borok, M. T. and V. V. Aleksandrov. Zav. Lab., 26, 214 (1960).
359. Dianov-Klokov, V. I. and V. A. Saltykova. Ibid., 26, 1124 (1960).
360. Etzel, H. W. J. Opt. Soc. Amer., 43, 87 (1953).
361. Lee, T. Appl. Spectr., 9, 5 (1955).
362. Gol'dreer, I. G. and M. L. Petrova. Technical Production Manual on Design and Development of Geophysical Instrumentation, No. 3, 59 (1959); Izmerit. Tekhn., No. 5, 57-59 (1960).
363. Symposium on Problems in Modern Physics, Series 3, No. 16, 5 (1951).
364. Electronics for Spectroscopists, Ed. by C. G. Cannon, London, 1960.
365. Frisch, S. E. Vestn. Leningr. Gosud., Univ., No. 8, 157 (1954).
366. Frisch, S. E. and V. A. Konovalov. Zh. Tekh. Fiz., 4, 523 (1934).

367. Frisch, S. E. and E. J. Schreider. Izv. Akad. Nauk SSSR, Phys. Ser., 13, 465 (1949).
368. Frisch, S. E. Vestn. Leningr. Gosud. Univ., No. 6, 26 (1950).
369. Frisch, S. E. Izv. Akad. Nauk SSSR, Phys. Ser., 18, 251 (1954).
370. Klauer, Fr. Ann. Phys., 20, 145 (1934).
371. Schreider, E. J. Zav. Lab., 18, 326 (1952).
372. Romand, J., G. Ballofet and B. Vodar. Spectrochim. Acta, 14, 454 (1959).
373. Frisch, S. E. Izv. Akad. Nauk SSSR, Phys. Ser., 4, 56 (1940).
374. Walsh, A. Spectrochim. Acta, 7, 108 (1955).
375. Filippov, A. N. Spectral Analysis and Its Uses. ONTI, 1937.
376. Fishman, I. S. Izv. Akad. Nauk SSSR, Phys. Ser., 9, 759 (1945).
377. Gatterer, A. Spectrochim. Acta, 1, 513 (1941).
378. Vinnichenko, E. N., A. N. Zaydel', and N. I. Kaliteyevskiy. Izv. Akad. Nauk SSSR, Phys. Ser., 18, 272 (1954); Vestn. Leningr. Gos. Univ., No. 11, 119 (1954).
379. Spektorov, L. A. and W. S. Engelst. Proc. of the 10th Conference on Spectroscopy, Lvov Univ. Press, Vol. II, 1958, p. 543.
380. Ivanov, N. P. Zh. Anal. Khimii, 15, 315 (1960).
381. Duffendack, O. S. and R. A. Wolfe. Proc. of the Sixth Summer Conf. on Spectr., p. 66, 1939.
382. Frisch, S. E., O. P. Bochkova and L. P. Razumovskaya. Optika i Spektroskopiya, 5, 624 (1958).
383. Bochkova, O. P. and L. P. Razumovskaya. Proc. of the 10th Conference on Spectroscopy. Lvov Univ. Press, Vol. II, 1958, p. 214.
384. Bochkova, O. P. and N. V. Chernysheva. Gazovaya Promyshl., No. 1, 49 (1961).
385. Chackett, K. F., F. A. Paneth and E. J. Wilson. J. Atm. Terrestr. Physics, 1, 49 (1950).
386. Chackett, K. F., F. A. Paneth, P. Reasbeck, and B. S. Wiborg. Nature, 168, 358 (1951).
387. Paneth, F. A. and K. Peters. Z. Phys. Chem., 134, 353 (1928).
388. Karlik, B. Sitzungsberichte der Akad. der Wiss. in Wien, Abt. IIa, 145, 145 (1936).
389. Mirtov, B. A. Usp. Fiz. Nauk, 63, 181 (1957).
390. Kaiser, H. Z. techn. Phys., 17, 227 (1936).
391. Rayskiy, S. M. Zav. Lab., 6, 265 (1937); 8, 124 (1939).
392. Berenshteyn, L. Ye., V. V. Nalimov and O. B. Fal'kova. Ibid., 27, 1254 (1961).
393. Nalimov, V. V. Ibid., 27, 1268 (1961).
394. Shayevich, A. B. Ibid., 27, 1276 (1961).

395. Nalimov, V. V. Treatment of Analytical Data by Methods of Mathematical Statistics. Fizmatgiz, 1960.
396. Malikov, M. F. Fundamentals of Metrology, Part 1: The Theory of Measurement. Publ. by the Measures and Gauges Committee of the USSR, 1949.
397. Nalimov, V. V., V. V. Nedler and N. P. Men'shova. Zav. Lab., 27, 861 (1961).
398. Kaiser, H. Spectrochim. Acta, 3, 40, (1948).
399. Mandelshtam, S. L. and V. V. Nedler. Optika i Spektroskopiya, 10, 390 (1961).
400. Mandelshtam, S. L. Izv. Akad. Nauk SSSR, Phys. Ser., 26, 848 (1962).
401. Brodskiy, A. I. Chemistry of Isotopes. Press of Acad. Sci. USSR, 1952.
402. Frisch, S. E. and V. I. Chernyayev. Dokl. Akad. Nauk SSSR, 3, 245 (1934).
403. Collie, J. N. and W. Ramsay. Proc. Roy. Soc., 59, 257 (1896).
404. Moureu, Ch. and A. Lepape. Compt. Rendus Acad. Sci., 152, 691 (1911); 153, 740 (1911); 174, 908 (1922); 183, 171 (1926).
405. Pentcheff, N. P. Ibid., 185, 511 (1927); 192, 691 (1931).
406. Heyes, J. Z. Phys. Chem., A172, 95 (1935)
407. Paneth, F. A. and K. Petters. Z. Phys. Chem., A134, 353 (1928).
408. Paneth, F. A. and P. L. Günther. Ibid., A173, 401 (1935).
409. Pettersson, H. Sitzungsberichte der Akad. der Wiss. in Wien, Abt. IIa, 142, 325 (1933).
410. Van Liempt, J. A. M. and S. H. R. Visser. Rec. Trav. Chim. Pays Bas, 53, 1084 (1934).
411. Monfils, A. and P. Posen. Rev. Univer. Mines, 6, 79 (1950).
412. Mellichamp, J. W., I. Adams and G. A. Wolff. Appl. Spectr., 15, 52 (1961).
413. Sventitskiy, N. S. Visual Methods of Emission Spectral Analysis. Fizmatgiz, Moscow, 1961.
414. Bochkova, O. P. and E. J. Schreider. Zav. Lab., 21, 311 (1955).
415. Bochdova, O. P. and E. J. Schreider. Vestn. Liningr. Gosud. Univ., No. 16, 57 (1956).
416. Dianov-Klokov, V. L. Zav. Lab., 22, 444 (1956).
417. Servigné, M., P. G. de Montgareuil and D. Dominé. Compt. Rendus Acad. Sci., 242, 2827 (1958).
418. Bochkova, O. P., L. P. Razumovskaya and V. G. Sagaydak. Kislorod (Oxygen), No. 4, 24 (1957).
419. Bochkova, O. P., L. P. Razumovskaya, S. E. Frisch, and N. V. Chernysheva. Optika i Spektroskopiya, 6, 818 (1959).
420. Bochkova, O. P., L. P. Razumovskaya and S. E. Frisch. Ibid., 5, 91 (1958).

421. Brit. Pat., 758,302, 1957.
422. U.S. Pat., 2,916,693, 1959.
423. Sagaydak, V. G. and A. V. Afonina. Zav. Lab., 27, 315 (1961).
424. Bochkova, O. P. and E. J. Schreider. Spectral Analysis of Gas Mixtures. Gostekhizdat, 1955.
425. Wignen, J. and A. Van-Tiggelen. Spectrochim. Acta, 4, 8 (1950).
426. Gatterer, A. Phys. Zeitschr., 33, 64 (1932).
427. Garger, K. S. Zh. Tekh. Fiz., 20, 1413 (1950).
428. Tombery. Experientia, 10, 388 (1954); Phys. Ber., 35, Heft 12 (1956).
429. Mirtov, B. A. and V. G. Istomin. Usp. Fiz. Nauk, 63, 227 (1957).
430. Mirtov, B. A. Gas Composition of the Earth's Atmosphere; Methods of Analyzing It. Press of Acad. Sci. USSR, 1961.
431. MacDonald, A. D. Proc. IRE, 47, 436 (1959).
432. Reasbach, P. and B. S. Wiborg. Rocket Exploration of the Upper Atmosphere. London, 1954, p. 158.
433. Bochkova, O. P. Izv. Akad. Nauk SSSR, Phys. Ser., 18, 252 (1954).
434. Musson-Genon, R. P. and E. Deschamps. Proc. of the Fourth Internat. Conf. on Ionisation Phenomena in Gases (Uppsala, 1959), Vol. I, 511 (1960), Amsterdam.
435. Morrison, J. 4th National Symposium on Vacuum Technology, Transactions, 1958, p.p. 100-103; cited from Ref. Zh. Fizika., No. 7, 16002 (1959).
436. Charles, D., R. J. Warnecke and J. C. Marchais. Video, 14, 274 (1959); cited from Ref. Zh. Fizika, 8, No. 20864 (1960).
437. Lilly, J. C. and T. F. Anderson. Report No. 299, National Research Council Committee on Aviation Medicine, 28, 11 (1944); cited from [438] and [439].
438. White, C. and W. R. Lovelage. Aviation Medicine, Selected Reviews, No. 25, 253 (1958).
439. White, C. Aviation Medicine, Selected Reviews, No. 25, 157 (1958).
440. White, C. S., W. H. Lockyear and L. L. Smith. J. Aviation Medicine, 26, 104 (1955).
441. White, C. S., L. C. Watkins and E. E. Fletcher. J. Aviation Medicine, 27, 332 (1956).
442. White, C. S., L. C. Watkins, E. E. Fletcher. J. Aviation Medicine, 27, 414 (1956).
443. White, C. S., L. C. Watkins and E. E. Fletcher. J. Aviation Medicine, 28, 406 (1957).
444. Nakagawa, K. J. Technol. Res., 4, 71 (1959).
445. Striganov, A. R. Usp. Fiz. Nauk, 58, 365 (1956).

446. Zaydel', A. N. Ibid., 68, 123 (1959).
447. Van Tiggelen, A. Bull. Soc. Chim. Belg., 55, 133 (1946).
448. Broida, H. P. and J. W. Moyer. J. Opt. Soc. Amer., 42, 37 (1952).
449. Broida, H. P. and G. H. Morgan. Anal. Chem., 24, 799 (1952).
450. Broida, H. P., H. J. Morowitz and M. J. Selgin. Res. Nat. Bur. Stand., 52, 293 (1954).
451. Oganov, M. N. and A. R. Striganov. Atom. Energ., 3, 112 (1957).
452. Oganov, M. N. and A. R. Striganov. Spectrochim. Acta, 13, 139 (1958).
453. Veynberg, G. V., A. N. Zaydel' and A. A. Petrov. Optika i Spektroskopiya, 1, 972 (1956).
454. Zaydel', A. N. and G. V. Ostrovskaya. Vestn. Leningr. Gosud. Univ., No. 16, 39 (1959).
455. Borgest, V. A. and A. N. Zaydel'. Optika i Spektroskopiya, 5, 686 (1958).
456. Brody, J. K. and F. S. Tomkins. Report at the 2nd Geneva Conference on Peaceful Uses of Atomic Energy, 15/P, 1945 (1958).
457. Artaud, J., J. Blaise and S. Gerstenkorn. Report at the 2nd Geneva Conference on Peaceful Uses of Atomic Energy, 15/P, 132 (1958).
458. Fergusson, R. E. and H. P. Broida. Anal. Chem., 28, 1436 (1956).
459. Zaydel', A. N. and G. V. Ostrovskaya. Optika i Spektroskopiya, 9, 137 (1960).
460. Hoch, M. and H. Weisser. Helv. Chim. Acta, 33, 2128 (1950).
461. Gorbunov, A. I. and P. A. Zagorets. Zh. Fiz. Khimii, 29, 1442 (1955).
462. Broida, H. P. and M. W. Chapman. Analyt. Chem., 30, 2049 (1958).
463. Zahn, H. Optik aller Wellenlange, Berlin, 1959, p. 58.
464. Zaydel', A. N., G. V. Ostrovskaya and A. A. Petrov. Optika i Spektroskopiya, 10, 673 (1961).
465. Urey, H. C., F. G. Brickwedde and G. M. Murphy. Phys. Rev., 40, 1 (1932).
466. Russel, B. J., J. P. Shelton and A. Walsh. Spectr. Acta, 8, 317 (1957).
467. Allan, J. E. Spectr. Acta, 14, 800 (1959).
468. Lockyer, R. and G. E. Hames. Analyst, 84, 385 (1959).
469. Zeil, W. Z. analyt. Chem., 170, 19 (1959).
470. Chulanovskiy, V. M. Introduction to Molecular Spectral Analysis. Gostekhizdat, 1950.

471. Lecomte, J. Infrared Radiation. Russian Edition, Fizmatgiz, 1958.
472. Dorgelo, H. B. Z. Phys., 34, 766 (1925).
473. Eckstein, L. Ann. Phys., 87, 1003 (1928).
474. Lecomte, J. Handbuch der Physik, B XXVI, 849, Springer, 1958.
475. Lecomte J. Bull. Soc. Chim., France, No. 7-8, 1080 (1959).
476. Rouir, E. V. Ind. Chim., Belge, 21, 221 (1956).
477. Williams, V. Z. Rev. Sci. Instrum., 19, 135 (1948); Russian translation, Usp. Fiz. Nauk, 37, 183 (1949).
478. Petrash, G. G. Optika i Spektroskopiya, 8, 122 (1960).
479. Brattain, R. R., R. S. Rasmussen and A. M. Cravath. J. Appl. Phys., 14, 418 (1943).
480. Coggeshall, N. D. and E. L. Saier. J. Appl. Phys., 17, 450 (1946).
481. Pemsler, J. P. and W. G. Planet. Spectr. Acta, 8, 302 (1956).
482. Herscher, L. W. Spectr. Acta, 14, 901 (1959).
483. Harrison, J. A. and C. D. Reid. J. Sci. Instrum., 36, 240 (1959).
484. Parsons, J. L., M. J. Irland and F. R. Bryan. J. Opt. Soc. Amer., 46, 164 (1956).
485. Pfund, A. H. Science, 90, 326 (1939); Fastie, W. G. and A. H. Pfund. J. Opt. Soc. Amer., 37, 762 (1947).
486. Veyngerov, M. L. Dokl. Akad. Nauk SSSR, 19, 687 (1938); Izv. Akad. Nauk SSSR, Phys. Ser., 4, 94 (1940).
487. Luft, K. F. Z. techn. Phys., 24, 97 (1943); Compt. Rendus Acad. Sci., 241, 482 (1956).
488. Veyngerov, M. L. Zav. Lab., 13, 426 (1947).
489. Luchin, S. M. Dokl. Akad. Nauk SSSR, 49, 428 (1945).
490. Sall', A. O. Optika i Spektroskopiya, 8, 135, 247, 571 (1960); Zh. Tekhn. Fiz., 26, 157 (1956).
491. Kluyver, J. C. and J. M. W. Milatz. Physica, 19, 401 (1953).
492. Kluyver, J. C. and E. W. Blokhuis. Physica, 20, 427 (1954).
493. Kluyver, J. C. Recueil Trav. Chim. Pays Bas, 74, 322 (1955).
494. Glasser, L. G. J. Opt. Soc. Amer., 45, 556 (1955); J. Electrochem. Soc., 97, 201 c (1950).
495. Broida, H. P., H. I. Schiff and T. M. Sugden. Nature, 185, 759 (1960).
496. Rasool, S. I., A. Vassy and B. Leroy. Ann. de Geophys., 12, 239 (1956).
497. Renzetti, N. A. Anal. Chem., 29, 869 (1957).
498. Chubb, T. A. and H. Friedman. Spectrochim. Acta, 8, 121 (1956).
499. Garton, W. R. S., M. S. W. Webb and P. C. Wildy. J. Sci. Instrum., 34, 496 (1957).
500. Prugger, H. and W. Ulmer. Z. angew. Physik, 11, 467 (1959).

501. Gerlovin, Ya. I. Optika i Spektroskopiya, 7, 571 (1959).
502. Handbook on Dosimetric, Radiometric and Electronic Physical Instruments, Counters, Scintillators and Photomultipliers. Published by the Central Administration for use of Atomic Energy, Council of the Ministers of the USSR, Moscow, 1959.

INDEX

Printed in Dunstable, United Kingdom

91343001R00185